Mini & Major Projects in

Instrumentation
for
Engineering Students

SA Khan

PUSTAK MAHAL®

Publishers
Pustak Mahal®

Administrative office and sale centre

J-3/16 , Daryaganj, New Delhi-110002
☎ 23276539, 23272783, 23272784 • *Fax:* 011-23260518
E-mail: info@pustakmahal.com • *Website:* www.pustakmahal.com

Branches
Bengaluru: ☎ 080-22234025 • *Telefax:* 080-22240209
E-mail: pustakmahalblr@gmail.com
Mumbai: ☎ 022-22010941, 022-22053387
E-mail: unicornbooksmumbai@gmail.com

ISBN 978-81-223-1543-1

Edition: 2018

Printed at : **Radha Offset, Delhi**

PREFACE

This is a book on Instrumentation Projects for engineering students. Instrumentation technology covers a wide range of subjects—physics, chemistry, material sciences, electronics, mechanical engineering and even embraces nanotechnology.

The ever-growing range of sensors, transducers and sophisticated semiconductor devices offer new scope for development, while the combination of biology with electronics is now a distinct discipline termed bionics.

This book is divided into four sections. The first part covers various experimental techniques, which are useful in the study and development of various instruments. The second part relates to hardware and chemicals which are useful in the development of projects. The third and fourth parts are devoted to mini and major projects for students.

In the selection of projects, the approach has been to cover a wide range of components and apply them in developing the projects. Some examples are optical fibres, thermoelectric modules, muscle wires, etc.

In the description of projects, spoon-feeding has been avoided wherever possible, leaving the students to use their own ingenuity in attending minor details. All in all, the projects are aimed at engineering graduates rather than amateurs and hobbyists. As such, most of the projects can be implemented using different approaches and using alternate hardware.

The students are encouraged to implement the projects using this line of action.

— SA Khan

Contents

PART I: EXPERIMENTAL TECHNIQUES

PART II: COMPONENTS AND SPECIAL CHEMICALS

PART III: MINI PROJECTS IN INSTRUMENTATION

PART IV: MAJOR PROJECTS IN INSTRUMENTATION

PART I

EXPERIMENTAL TECHNIQUES

Sources of Power : DC Supplies

In most of the circuits employing semiconductor devices, DC power is required. This may be either through batteries or by conversion of AC into DC. We will consider both options in this and the next chapter.

Conversion of AC into DC

In the projects described in this book (with one exception), DC voltages are not more than 15V. Since the main line voltages in India are around 220-230 VAC, it is desirable to step this down to the required levels before conversion into DC. Stepping down AC voltages is easily done through the use of appropriate transformers. We will be discussing some useful circuits for achieving this objective.

DC Power Supply Process

The process of developing a DC power supply is as follows:

(a) Step down the mains voltage from 230 V AC to the required level.

(b) Convert the AC into DC by use of rectifiers. This leaves an AC ripple which has to be smoothed out.

(c) Remove the AC ripple through the use of capacitors.

(d) Finally, obtain the required steady DC voltage by the use of semiconductor devices.

While the first three steps are relatively straight-forward, we have three options for the final stage:

i. Use Zener diodes of appropriate voltage and power to give the desired DC output. See the chapter on Zener diodes for further information.

ii. Use discrete transistors in combination with Zener diodes to provide the required voltage output.

iii. Use special ICs to carry out the required final operation.

We will discuss all the three options hereunder.

Zener Controlled Power Supply

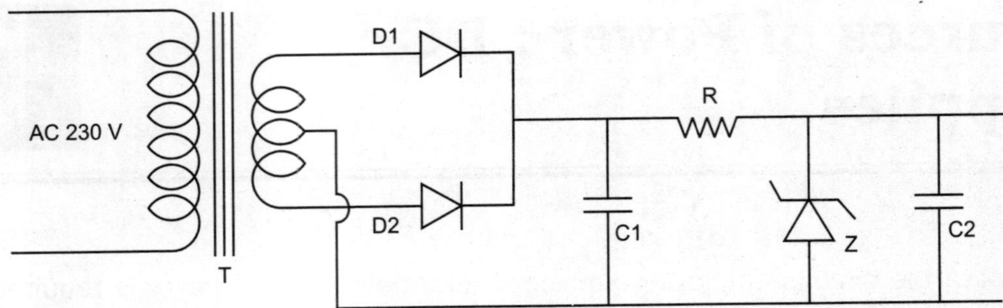

Figure 1.1 *Zener controlled power supply*

In this circuit, a step-down transformer with centre-tapped secondary provides the low voltage, which is rectified by dual rectifiers D1 and D2. A filter capacitor removes the AC ripple from the DC.

A series resistance connects the Zener diode, which holds the voltage at its rating.

If a centre-tapped transformer is to be avoided, then a bridge rectifier can be used as shown below. This arrangement uses four rectifier diodes instead of two, wherein a centre-tapped transformer is not required.

A commonly used circuit where higher currents are required than what can be handled by a single Zener, is given in Figure 1.3. This arrangement provides a higher current output.

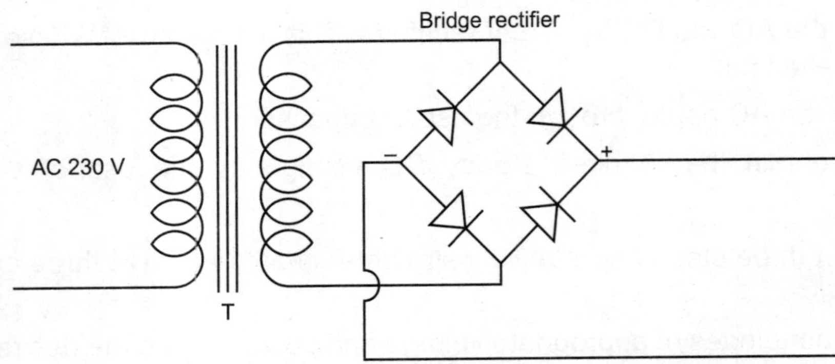

Figure 1.2 *Conversion of AC into DC through a bridge rectifier*

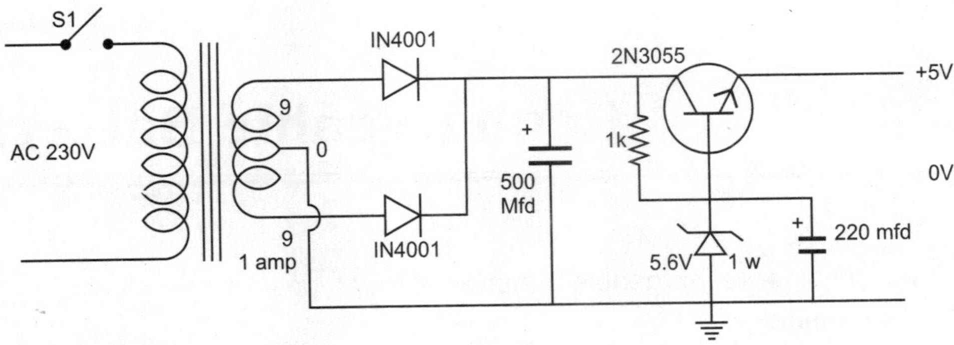

Figure 1.3 *DC power supply with transistor regulation*

Although the circuit is arranged to provide a 5 V supply, other voltages can be obtained by a suitable choice of the transformer and the Zener diode.

Regulated DC supplies are very easily fabricated using voltage regulator ICs, of which, a large variety is available for different voltages. The circuit below gives a fixed voltage of five volts through the use of a single IC. Voltage regulator ICs come in various voltage outputs; up to 24 V being quite common.

The circuit given below uses a minimum of components.

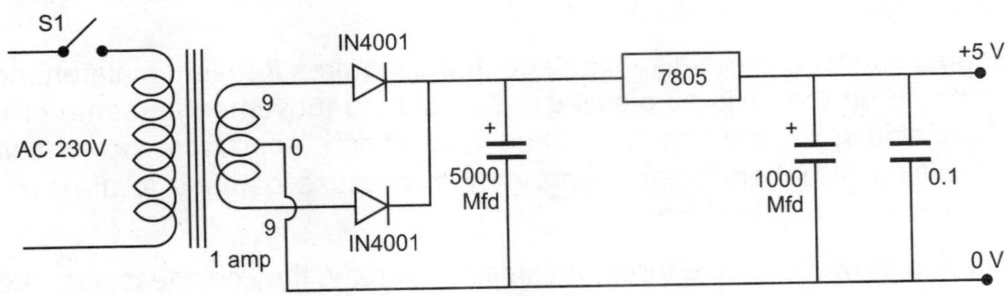

Figure 1.4 *IC regulated power supply*

So far, we have discussed power supplies with fixed voltages. A variable voltage power supply can be fabricated using the same basic components, with slight variations. The circuit shown below provides a voltage variable up to 12 V.

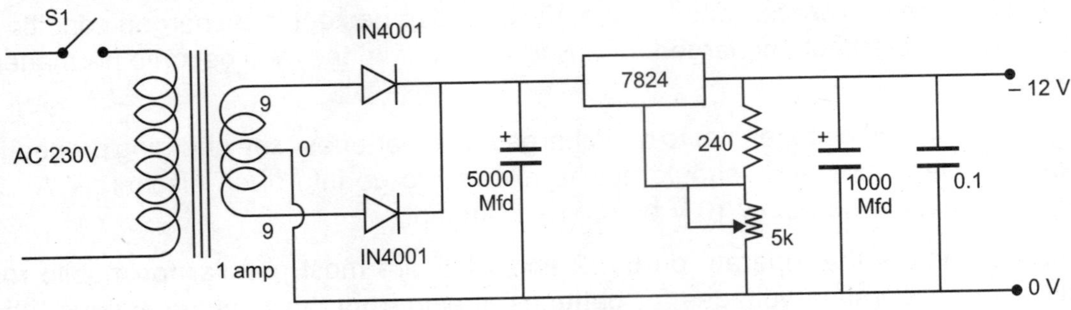

Figure 1.5 *Variable voltage using IC regulator*

2 Rechargeable Batteries

Here, we will consider rechargeable batteries of four types:

(a) Lead-acid batteries

(b) Nickel–cadmium cells

(c) Nickel metal hydride

(d) Lithium ion batteries

Lead-acid Batteries

These are cost-effective, easily available and provide adequate power for most applications. They come in two types—

i. Unsealed batteries requiring periodic maintenance.

ii. Sealed maintenance-free batteries.

The unsealed variety has the disadvantage that it requires frequent maintenance in the form of topping up the cells by distilled water. During movement, if some of the cells leak and the acid spills, then fresh acid has to be added. Apart from the inconvenience involved, spilling of battery components may also cause damage to the surrounding parts, as the acid is highly corrosive.

Such batteries, however, have three advantages. Firstly, they cost less than the sealed variety; secondly they are less sensitive to overcharging (which frequently damages the sealed types); and thirdly, if they are not required for an extended period, they can be safely stored by removing the electrolyte and washing the plates. When required again, they need to be filled with fresh acid and recharged.

Since sealed batteries are maintenance-free, they have a decided advantage over the unsealed variety. However, once in use, they have to be kept in a charged condition. If such batteries are kept uncharged for any length of time, they will become permanently useless.

Ample care should be taken not to overcharge these batteries (*see* charging procedures below). While in use, they should not be allowed to go into deep discharge. A 12 V battery should not go below 10 V before re-charging.

Since DC motors that operate on 6, 12 and 24 V are most popular for mobile robot applications, operating voltages for batteries is also from the above-mentioned three voltage ratings.

For a given power requirement, the lower the battery voltage, the higher is the current that has to be provided. This also requires semiconductor devices used in the drive circuits to be able to handle higher currents for motor controls. Hence in the choice of batteries, both the voltage and current capacities have to be taken into consideration.

Another battery specification is the ampere-hour (AH) capacity. This specifies how much charge the battery can store, which depends on the size and the number of such plates in each cell of the battery.

To illustrate this, consider a battery with a capacity of 10 AH. Theoretically, such a battery should provide 1 amp of current for a period of 10 hours, 2 amps for five hours and up to 10 amps for an hour.

Given examples will clarify these points.

We will consider a permanent magnet DC motor whose power output is 100 watts. Assuming the motor efficiency to be 60 per cent, we will require an input power of 166 watts to the motor.

We will consider here three DC motors operating at different voltages:

Operation voltage (V)	24	12	6
Full load current (Amps)	7	3.5	1.75

Now consider batteries with different capacities:

Battery capacity in Amp-hours (AH)	10	20	40

Operating time (in hours) before re-charge (assuming 70% discharge):

6V battery	1	2	4
12V battery	2	4	8
24V battery	4	8	16

For a given voltage, the higher the capacity of the batteries (in AH) is, the higher is the cost and larger is the weight and size. It should be noted that a 6 V motor will consume four times current than a motor operating at 24 V, of the same power and efficiency. Hence, the power wiring and control devices must be suitably selected with this factor in mind.

Rechargeable batteries are available in various voltage and current capacities, the 6 V and the 12 V varieties being commonly available. Such batteries are assembled from basic 2 V units, connected in series. For a 12 V requirement, six such 2 V cells are connected in series and integrated in a single package. While this arrangement is very convenient, it has one drawback. If one of the cells gets damaged, the entire battery has to be replaced. On the other hand, if 2 V cells are used separately, only the damaged cells need to be replaced.

By connecting such batteries of equal voltage parallelly, the current capacity of the assembly can be increased. With such series-parallel arrangements, we can obtain a wide range of voltages and currents for different applications. The figures given below will illustrate this.

Since the batteries need to be charged from time to time, suitable arrangement must be incorporated in the mobile robots for this purpose. The battery charger can be installed within the robot; with the corresponding increase in the weight of the robot; or it can be an external one—with a suitable connector provided on the robot for the charging operation.

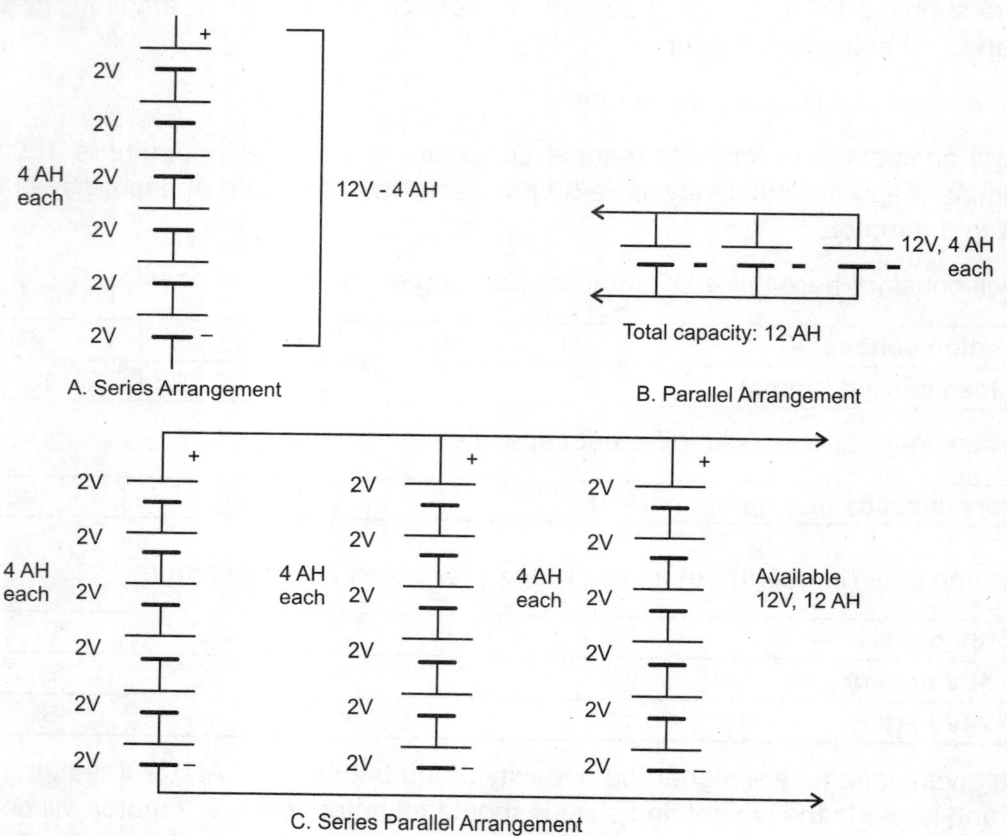

Figure 2.1 *Batteries in series and parallel*

Charging the batteries has to be in accordance with the manufacturer's guidelines. Charging at the currents higher than that specified by the manufacturer can seriously reduce the battery life or even damage it permanently, due to overheating of the battery plates.

Construction of Lead-Acid Cells

Each of the cells in a lead-acid battery consists of two plates—one made of lead and the other of lead oxide. The plates are fabricated in a grid form as shown below:

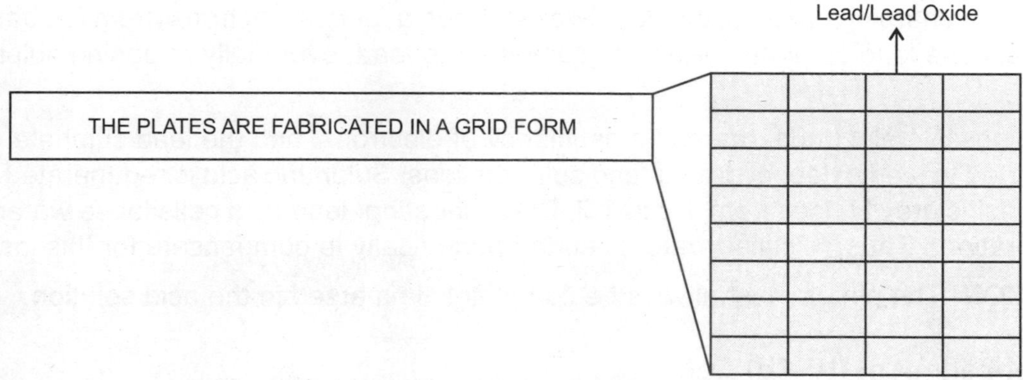

Figure 2.2 *Cross-section of battery plates*

The plates are inserted in a non-conductive container like Bakelite and insulated from each other using a rubber separator.

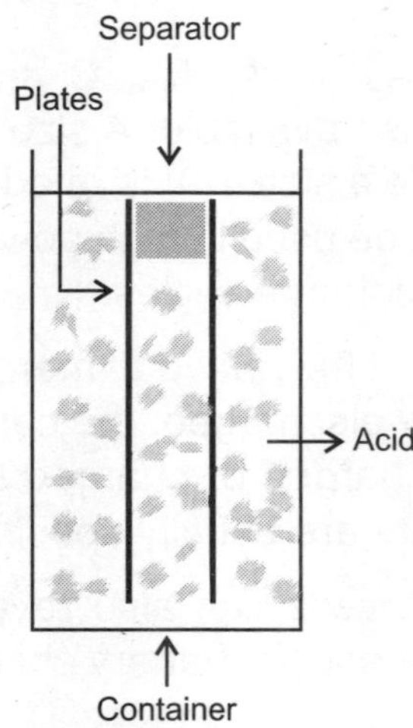

Figure 2.3 *Cross-section of lead-acid battery*

The container is filled with dilute sulphuric acid so that the plates are completely immersed. The cell needs to be 'charged' before use. Initial charging takes 36 to 48 hours at currents larger than 1 amp, depending upon the size of the cell. A fully-charged cell delivers current at two volts. The electrode connected to the lead plate is the negative pole of the cell, while that connected to the lead oxide plate is the positive one.

During discharge, the lead (-ve) electrode loses electrons; the resulting positive lead ions combine with the negative sulphate ions forming lead sulphate (Pb SO4) and releasing hydrogen from the sulphuric acid.

The lead oxide positive electrode gains electrons, the lead combining with sulphate ions to form lead sulphate releasing oxygen, which combines with hydrogen to form water. The sulphuric acid becomes weaker, with its specific gravity decreasing from 1.3 to 1.2. Both the plates now get converted to lead sulphate.

During the charging process, the negative electrode acquires electrons (from the battery charger); the lead sulphate getting re-converted to lead, eventually releasing sulphate ions.

At the positive electrode, there is a deficiency of electrons, and the lead sulphate gets converted to lead oxide, hydrogen and sulphate ions. Sulphuric acid is regenerated and the specific gravity rises from 1.2 to 1.3. Due to heating, lead acid cells loose water (by evaporation). Thus, distilled water is added periodically to compensate for this loss.

CAUTION: The plates must always be completely immersed in the acid solution.

Nickel-cadmium (Ni-Cd) Cells

For their positive electrode, Nickel–cadmium cells have graphite and nickel hydroxide; while the negative electrode is composed of cadmium and iron oxide. The electrolyte is potassium hydroxide. They deliver a constant voltage of 1.2 V over 90 % of their discharge cycle.

These cells come in various sizes, AAA, AA, C and D being the most common. Rechargeable 9 V packages are also available. A size 'D' has got a current capacity of 1200 mAh (milliampere-hour), while a size 'AA' is rated at 500-700 mAh. They have low internal impedance and hence can be used in high power/pulsed applications. They are commonly rated for 500 charge-discharge cycles.

A major drawback of the cells is that they have a 'memory effect'. When a Ni-Cd cell is recharged before being completely discharged, the capacity of the cell decreases. The cell can also be damaged if overcharged or charged too rapidly. Ideal charging rates for cells from various manufacturers are usually specified on them.

Apart from the memory effect, the cells can also reverse their polarity under certain conditions. For charging such cells, special battery chargers are available which should be used.

Nickel-metal Hydride (NiMH)

Amongst the re-chargeable types of batteries, NiMH ones are the most efficient and are available in all standard sizes. They have low internal resistance, hence can supply large currents. These batteries can be recharged about 500 times and are ideal for robotic applications.

Special chargers should be used to charge these batteries. Some chargers are available which provide the option to charge either Ni-Cd or NiMH types.

An advantage that such types have over Ni-Cd is that they do not have the memory effect. If required, they can be charged before completely discharging or can be given a short duration charge and put to use. This is a major advantage as these batteries do not retain the charge for very long, when idle. Even when not being used, frequent recharging is advisable.

These batteries supply current at 1.2V similar to the Ni-Cd variety. In actual use, they tend to get hot and this should be kept in mind when placing these near temperature sensitive devices. Provide sufficient cooling (air circulation) or thermal insulation for other devices in the vicinity.

Lithium-ion Batteries

Lithium-ion batteries have become very popular since they are widely used in cell phones, laptops and iPads. They are the most energetic rechargeable batteries available at present. The electrodes of these batteries are lithium and carbon. Lithium is a light-weight element. In the periodic table of elements, it is next to hydrogen and helium, both gases. Hence, lithium with atomic number 3 is the first solid in the periodic table.

Compared with other types of batteries, Lithium ion batteries have six times the capacity of lead acid type and 50 per cent more than NiMH type. Another advantage is that they do not have any memory effects, unlike the Ni Cd type, and can be recharged hundreds of time.

The disadvantages of these batteries include their sensitivity to high temperatures, which make them accident-prone. Their life is also short—two to three years (whether they are being used or not). An important precaution is that they should not be completely discharged otherwise they will be ruined.

3 Basic Electroplating Techniques

Copper Plating

A simple method of plating iron with copper is described below. This serves a double purpose:

(a) Prevents iron from rusting

(b) Provides better electrical contact between wires and terminals.

For electroplating, we need a source of electricity. One of the easiest to fabricate is the Daniell Cell. This is made up of two electrolytes—sulphuric acid and copper sulphate, and two electrodes—one of copper and the other of zinc. See Figure 3.1 in coloured illustrations on pages 177–184.

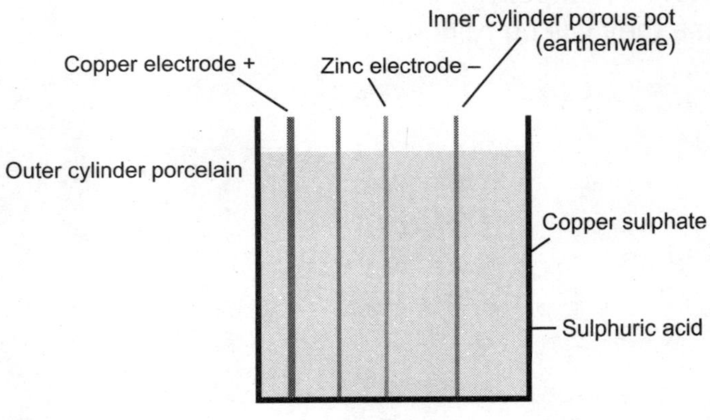

Figure 3.1

The electrolytes begin to dissolve zinc and copper producing positively-charged ions. With the ions migrating from zinc and copper plates, the free electrons are left behind. As zinc dissolves faster than copper, more electrons are created here, making the zinc electrode more negative than copper. The voltage thus produced is 1.1V.

An arrangement for plating copper on iron is shown in Figure 3.2 in coloured illustrations.

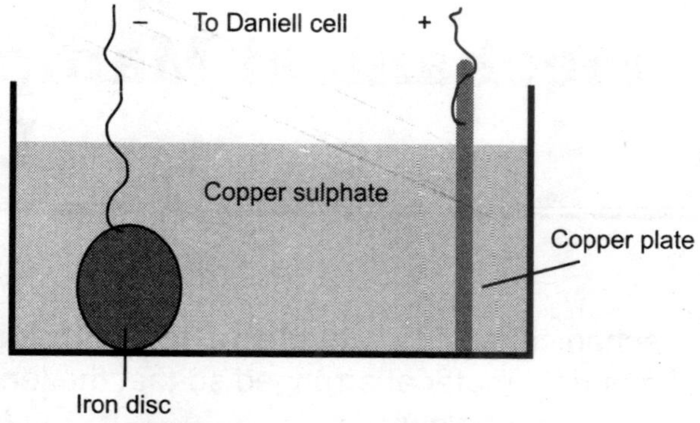

Electroplating bath: Copper gets deposited from the plate to the iron disc.

Figure 3.2

Silver Plating

Silver plating is best done on a copper base. The principle is the same as that of copper plating described above, using a Daniell Cell as the source of power. However, anode should be of pure silver and electrolyte should be a silver salt. Silver nitrate can also be used but best results are obtained by using silver cyanide as the electrolyte.

Silver cyanide is highly poisonous and should be handled with extreme care. When an iron component has to be plated with silver, it should first be plated with copper, cleaned and then silver-plated. Obviously, different baths should be used for copper and silver plating.

Gold Plating

Gold can be plated on copper and silver. The anode should be of pure gold while the electrolyte should be gold chloride. High quality connectors have their pins gold plated since gold does not tarnish easily. Whatever plating has to be done, the object to be plated should be thoroughly clean, free of grease and other impurities.

4 Mechanical Manipulators—Lazy Tong

The 'Lazy Tong' is a mechanical device with interesting applications. Basically, it is a set of strips or tubes of any rigid material arranged so that the length of the unit can be extended or retracted. See figure below:

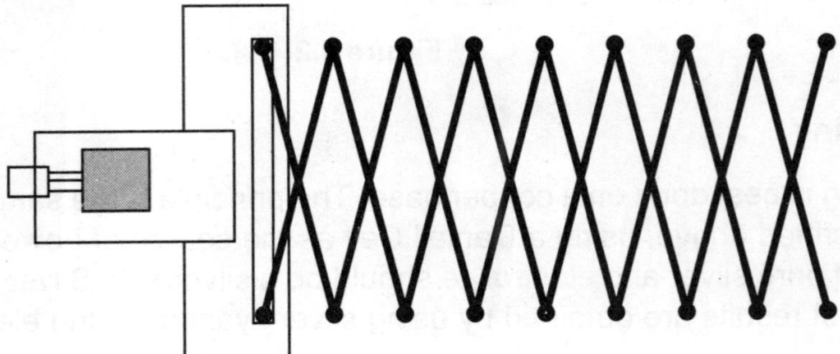

Figure 4.1 *Lazy tong compacted*

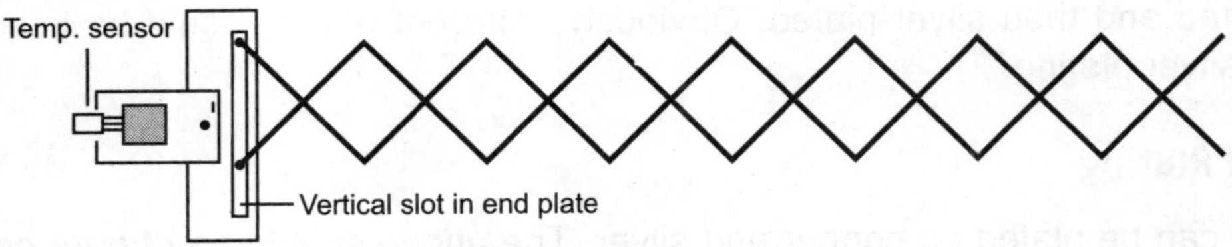

Figure 4.2 *FigureLazy tong extended*

The cross-links are pivoted at their centre and the ends as shown below:

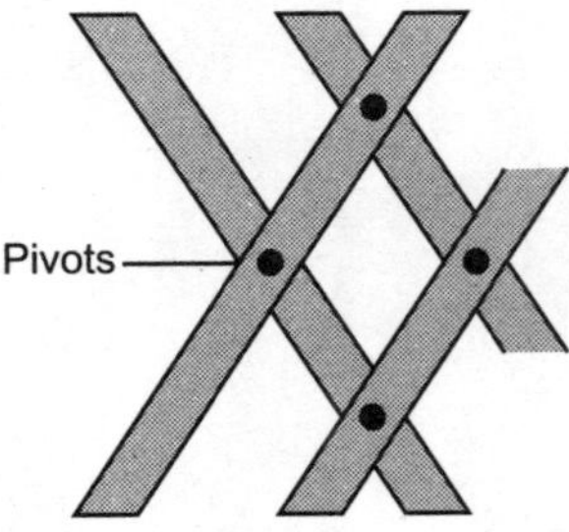

Figure 4.3 *The pivot points of the linkages*

The operating distance will depend upon how much the lazy tong can be extended. This, in turn, depends upon the distance between the pivots and number of links in the complete mechanism.

The figures below show the distance between two links when compacted and fully extended.

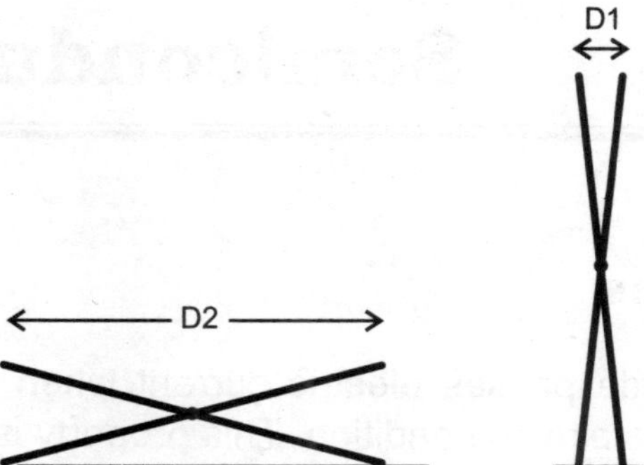

Figure 4.4 *Consecutive sections of the tong when compacted and extended*

The extension of the lazy tong can be determined if we measure the two distances D1 and D2. Then the extension will be:

N (D2 – D1), where N is the number of links in the mechanism.

Provision has to be made to extend linkages and withdraw them, whenever desired. This can be achieved by operating the complete tong through a screw which in turn is rotated by an electric motor.

Referring to Figure 4.4, the drive screw is shown connected to the motor by a set of gears. The screw has two sets of threads—the top part has left hand threads while the lower part has right hand threads. The linkages are connected to the screw through corresponding nuts. Thus, when the motor rotates, due to the reverse thread arrangement, one of the nut traverses down while the other one moves up, resulting in the ends of the tong shortening or lengthening.

The motor shown is a permanent magnet DC motor, whose direction of rotation can be changed very easily.

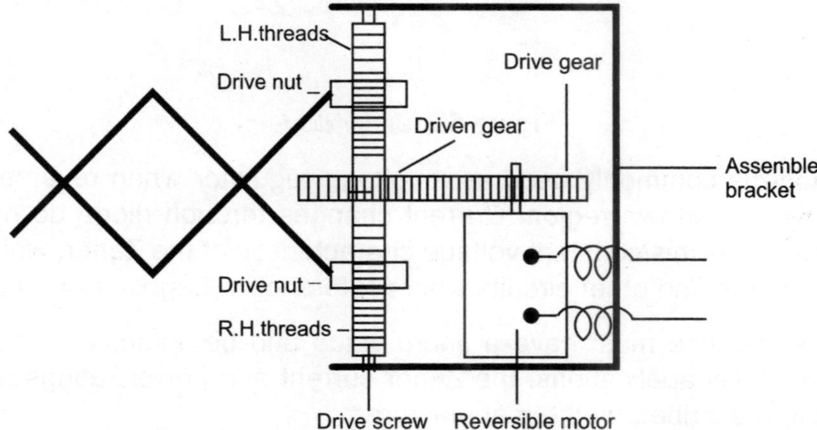

Figure 4.5 *Arrangement for driving the lazy tong using a reversible motor*

5 Semiconductor Basics

Semiconductor Diodes

A silicon p-n junction diode passes electric current when forward biased, while no current flows is the reverse biased condition. This property is utilised in the application of p-n diodes as rectifiers are commonly used in power supplies to convert AC to DC.

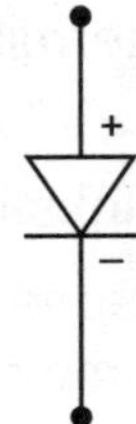

Figure 5.1 *P-N junction diode*

A Zener diode is basically a p-n junction diode used in the reversed-bias condition. No conduction takes place until reverse voltage exceeds the pre-determined value, or reverse breakdown voltage (known as the Zener voltage) of the particular diode, determined by its manufacturing process.

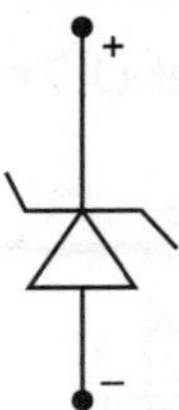

Figure 5.2 *Zener diode*

The Zener diode is commonly used as a voltage regulator when reverse-biased and operates in the breakdown region. Current changes through diode do not affect the voltage across it. It is this constant voltage characteristic of the Zener, which is utilised in voltage regulators and other circuits where reference voltages are required.

The breakdown voltage must have a sharp value and be insensitive to temperature changes. In practical applications, the Zener current and power ratings must be kept within the limits prescribed by the manufacturer.

Zener diodes come in ratings from 1 volt to several hundred volts. Such diodes, connected in series, provide higher voltage devices. See figure below:

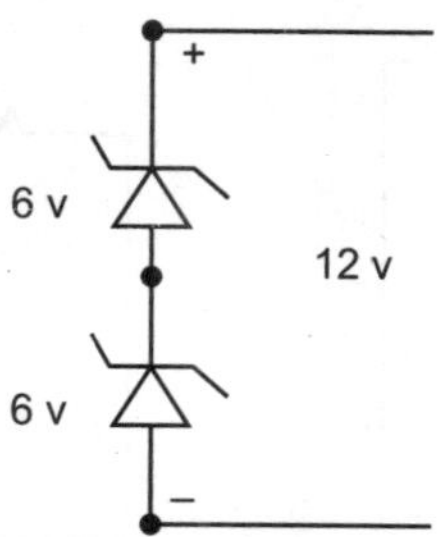

Figure 5.3 *Zener diodes in series*

Transistors

The transistor was invented in 1949 and since then, it has revolutionized the electronics industry. As technology progressed, many transistors were packaged into a single device–the integrated circuit (or IC, as it is commonly known).

The entire computer industry, communication networks and systems are based on such semiconductor devices; the latest entrant being the cell phone.

Currently, the most widely used transistor is the silicon one. These are of two types—NPN and PNP.

In practice, transistors can be used either as linear devices in signal amplifiers or as digital devices used in switching applications.

In linear devices, the output signal will be proportional to the input signal. For instance, when used in motor control circuits, the motor speed will be proportional to the control signal applied to the drive transistors. The transistors used in such applications can either be voltage amplifiers, which enhance the amplitude of the input signal voltage; or current amplifiers, which provide current amplification and are used in power systems. The voltage or current amplification available in any transistor circuit will depend on the circuit design parameters as well as the characteristics of the transistor itself.

In the case of transistors used in switching applications, the device acts as an on-off switch. In absence of an input signal, the transistor is in 'off' (non-conducting) state. When an appropriate signal is applied to the input, the transistor goes in 'on' state and passes a current that depends on the circuit configuration.

The following figures indicate the use of PNP and NPN transistors as linear devices or in switch-mode application. Note that the battery polarities are reversed in PNP devices as compared to the NPN ones.

The transistor in linear mode operation – as amplifier

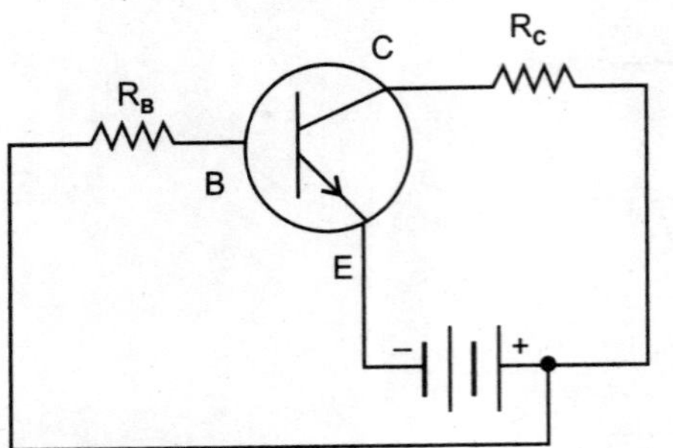

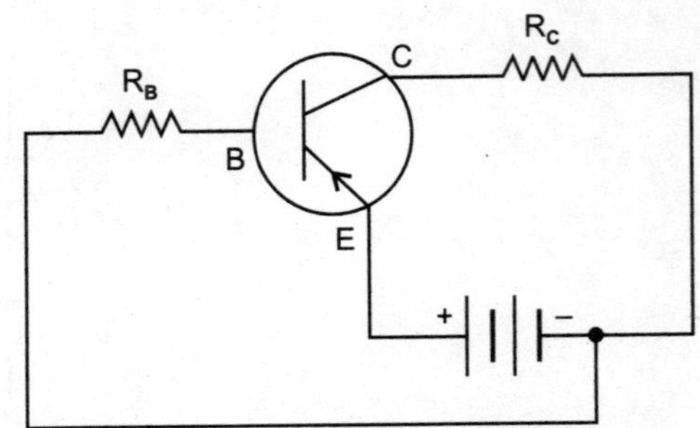

Figure 5.4 *Transistor types*

Note that the above two circuits are identical except for battery polarities.

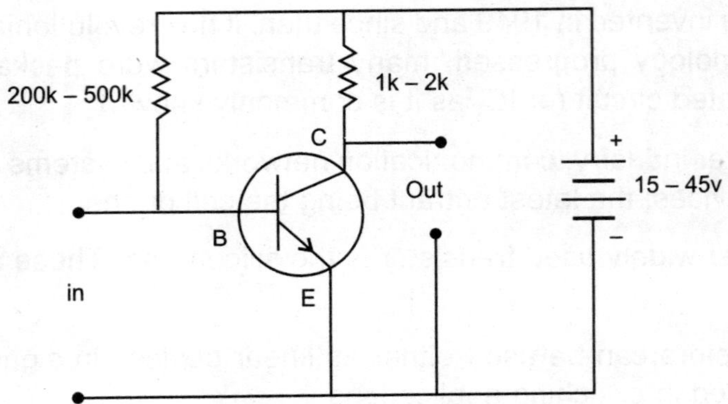

Figure 5.5 *NPN transistor amplifier*

$$\text{Current gain } (\beta) = \frac{\text{Collector current}}{\text{Base current}} = \frac{\Delta Ic}{\Delta Ib}$$

Most commonly, small signal transistors have β values ranging from 40–300 while power transistors have β values in the range of 10–30.

The Transistor as Switch

The common toggle switch based on mechanical action has serious limitations when used in high speed switching applications (*see* chapter on Mechanical switches). The silicon transistor serves as a useful alternative. The figure below shows how a transistor can be used in this mode.

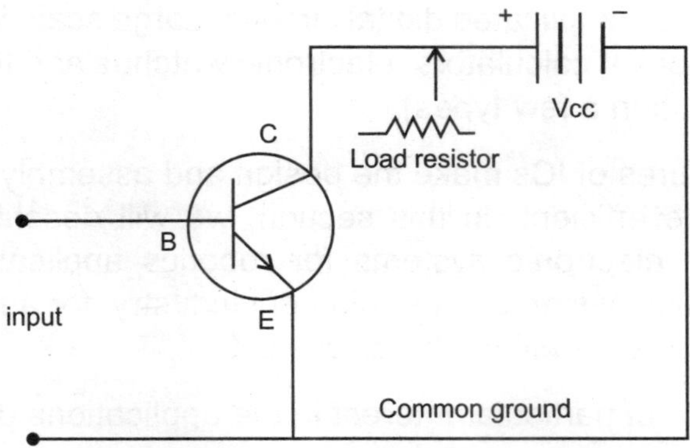

Figure 5.6 *The transistor as a switch*

When the base is open, there is no current in the collector-emitter circuit, and the collector voltage is Vc. In this state, the transistor is switched off. If a positive voltage is applied to the base, the transistor switches on and current flows in the C-E circuit, causing the collector voltage to drop to almost the ground level. In the case of a silicon transistor, this will be 0.6 V.

In practice, a load resistor should be connected in the collector circuit to limit the current through the transistor to a safe level and prevent the transistor from getting damaged. The base voltage required to switch on the transistor is typically 1 volt or more. Care should be taken not to exceed the maximum safe base-emitter voltage of the particular device being used. Consequently, if we apply a step voltage to the base, an inverted step voltage is obtained from the collector.

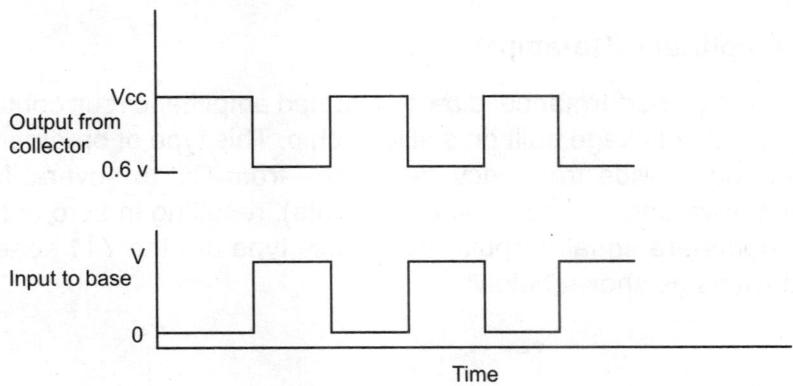

Figure 5.7 *Switching action of a transistor*

Integrated Circuits

Hundreds of transistors and diodes combined with resistors and capacitors assembled on a single silicon wafer form what are known as integrated circuits (ICs). They are available in various categories from simple logic circuits, memory circuits, voltage and operational amplifiers to complicated digital circuits. Large scale integrated circuits (LSI devices) are used in pocket calculators, electronic watches and the recently developed mobile phones (to mention a few types).

The standardized features of ICs make the design and assembly of electronic systems much easier and more efficient. In this section, we will describe some of the more common ICs used in electronic systems for robotics applications. ICs specifically meant for communication systems, entertainment industry, for automobile applications, medical instrumentation etc., will not be discussed.

The following types are of particular interest in the applications discussed in this book:

(a) Operational amplifiers (op-amps)

(b) Digital gates

(c) Digital Counters and shift registers

(d) Multivibrators including mono-stable and astable

(e) Phase-locked loops (PLLs)

(f) Timers

These ICs are broadly divided into two groups:

i. Analogue devices

ii. Digital devices

While op-amps, PLLs and timers are classified as analogue devices, digital gates, counters, shift registers and multivibrators are digital devices.

Operational Amplifiers (Op-amps)

An op-amp IC is a high performance, directly coupled amplifier circuit containing several transistors in a single package built on a silicon chip. This type of op-amp is a high-gain signal amplifier with a wide frequency response—from DC to several MHz. It has a differential input (inverting and non-inverting inputs), resulting in zero output voltage—when the two inputs are equal. Popular ICs of this type are the 741 series. These are five-terminal devices as shown below:

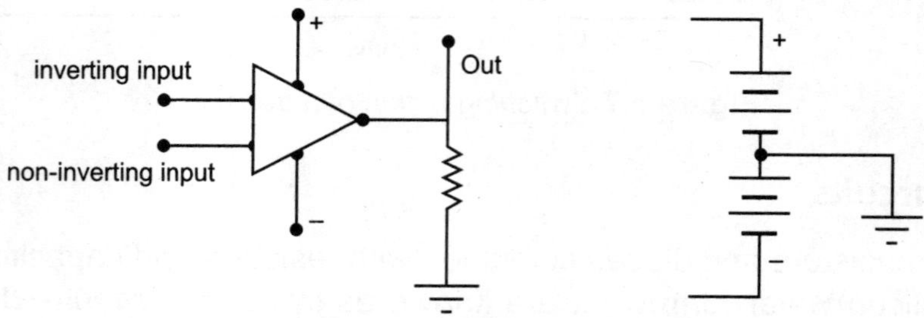

Figure 5.8 *An operational amplifier*

A bipolar power supply is required for their operation providing +ve and –ve voltages, with the centre connected to the circuit ground.

Digital Integrated Circuits

Some commonly used digital ICs of the TTL (transistor-transistor logic) are shown below.

Digital Gate Type 7400

input 1, 2 — output

A NAND gate
with truth table at right

Truth Table

Input 1.	Input 2.	Output
0	0	1
0	1	1
1	1	1
1	1	1

7400 quad NAND gate package

14 13 12 11 10 9 8

Vcc

Gnd

1 2 3 4 5 6 7

Figure 5.9 *A NAND gate IC*

Note: While the 7400 IC NAND gates have 2 inputs each, ICs with more inputs are also available. For example, IC 7422 has two gates with 4 inputs each, while IC 7430 has one gate with 8 inputs.

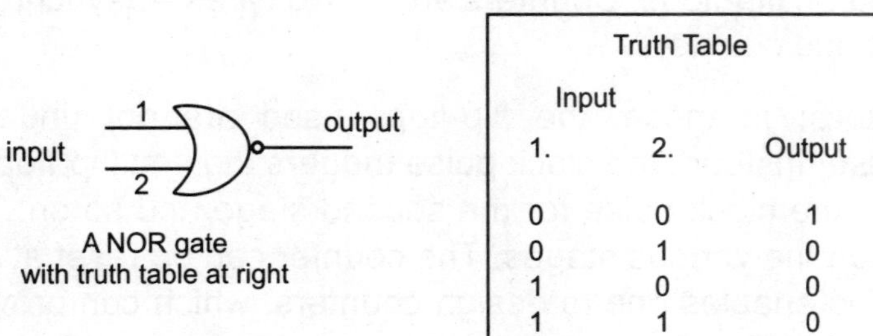

Truth Table

Input 1.	Input 2.	Output
0	0	1
0	1	0
1	0	0
1	1	0

IC type 7402 contains 4 dual input NOR gates in a single package. The loading factors are identical with7400.

Figure 5.10 *Type 7402 NOR gate*

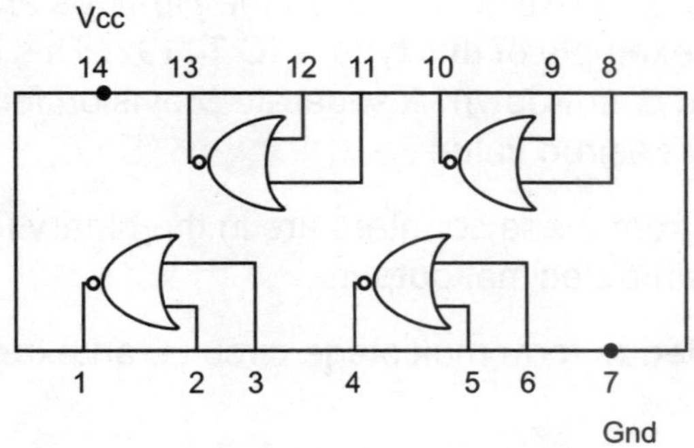

Figure 5.11 *IC7402*

Inverter Gates

IC type 7404 contains 6 inverter gates

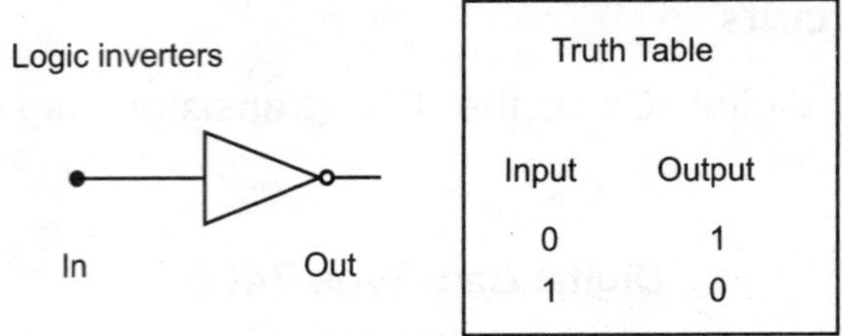

IC type 7404 contains 6 inverter gates

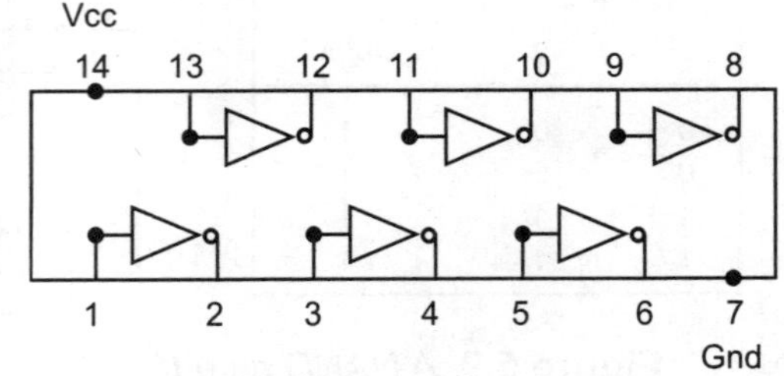

Figure 5.12 *IC 7404 inverter and truth table*

Digital Counters

Most digital counters are based on flip-flops. Counters are of two types—asynchronous (ripple) counters and synchronous counters.

Asynchronous/ripple counters: In these, the flip-flops used are not under the command of a single clock pulse. Instead, the clock pulse triggers the first flip-flop. The output of this flip-flop provides the clock pulse for the second stage and so on. Thus, the input trigger 'ripples' through the various stages. The counter can be reset at some selected state of the count. This enables one to design counters, which can provide a wide range of count outputs. Examples of asynchronous counters based on ICs are the BCD counter type 7490 and the modulo-16 type 7493.

Synchronous counters: In these counters, all the flip-flops are under the command of a single clock pulse. An example of this type is IC 74192. This device can count in both directions—count-up and count-down. A separate provision for data inputs enables the counter to be rest to any desired value.

Since the count outputs from these counters are in the binary form, a suitable decoder-driver is required to obtain a decimal output.

Counters can be cascaded to form multistage circuits, an example of which is given in the figure below.

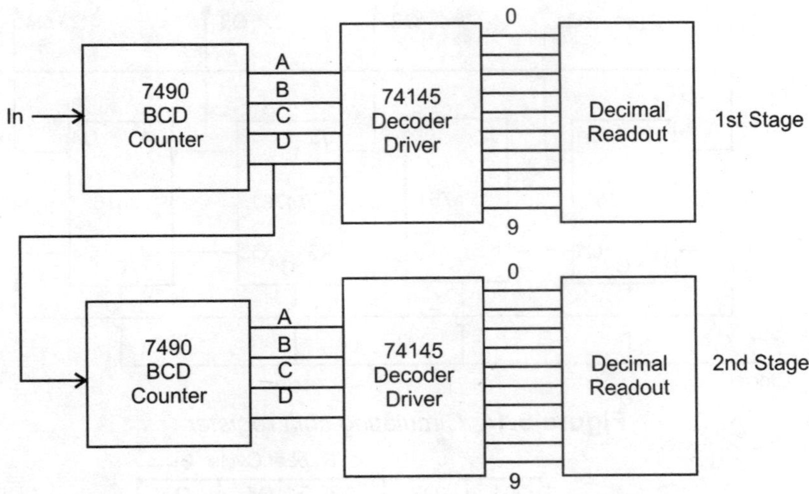

Figure 5.13 *Two stage decade counter*

Decoder-drivers

Decoder-driver ICs used with counters serve the following purposes:

(a) Convert the binary at the BCD input into ten or sixteen discrete outputs—only one of them being active'

(b) Provide or sink sufficient current to drive a suitable display or operate a relay-drive circuit

The outputs of a decoder can either be low or high, on being activated.Examples of the former type are 7441, 74145 and 74154 while the latter type includes 7447 and a binary-7 segment decoder.

Shift-registers

Two types of shift register chains used in the models are described below:

Circulating shift-register: A circul ating shift register is required in models where an LED display has to run continuously. The Q and the $\bar{Q}$ outputs of the last flip-flop are connected to the 'J' and 'K' inputs respectively, of the first one.

This results in circulating the 'high' Q output from one flip-flop to the next at each clock-pulse round the shift register chain. This is illustrated below in Figure 5.14, which shows a four-state shift register together with the truth table (shown in Figure 5.15).

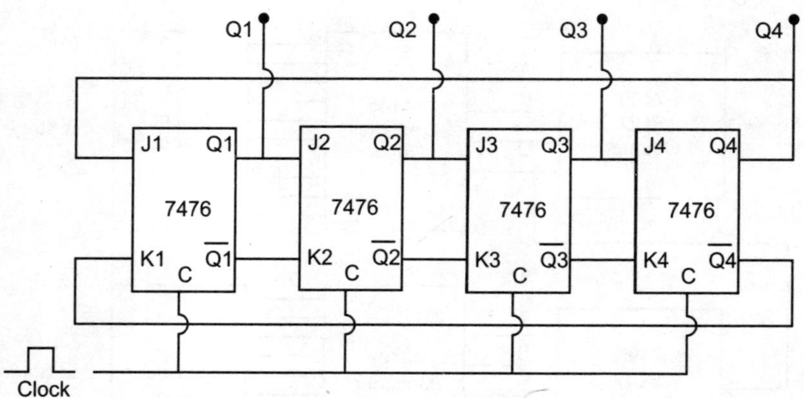

Figure 5.14 *Circulating shift register*

Repeat Cycle ←

	Q1	Q2	Q3	Q4	Q1
Preset →	1	0	0	0	1
	0	1	0	0	0
	0	0	1	0	0
	0	0	0	1	0

Figure 5.15 *Truth table of circulating shift register*

Non-circulating shift registers: When the 'J' and 'K' inputs of the first flip-flop are suitably fixed-biased, the high 'Q' progressively advance at each clock pulse, resulting in the low 'Q' being finally eliminated. Figure 5.16 and 5.17 will make this clear.

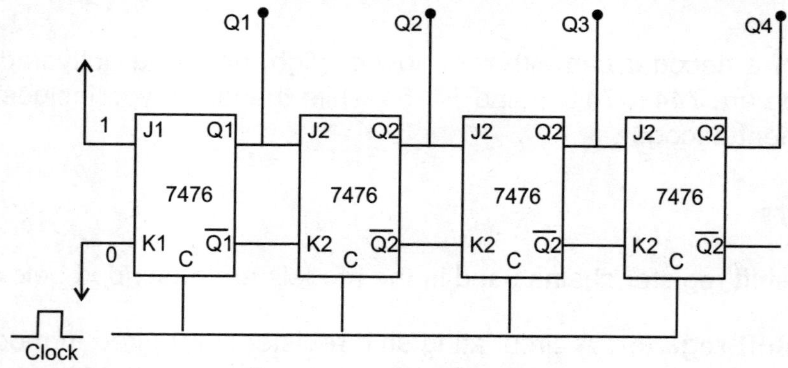

Figure 5.16 *Four-stage non-circulating shift register*

Truth Table	Q1	Q2	Q3	Q4
RST	0	0	0	0
CP1	1	0	0	0
CP2	1	1	0	0
CP3	1	1	1	0
CP4	1	1	1	1

Figure 5.17 *Truth table for non-circulating shift register*

Multi-stage shift registers are available in single packages, but without Q outputs available for each stage.

The TTL ICs commonly used for shift registers covered here are 7473 and 7476 types. They can provide enough current for driving one LED directly. Buffered outputs can be used when a larger fan-out is required.

Astable and Monostable Devices

IC type 555: A popular IC is the type 555 which is a very versatile device. It can be used as a timer, a monostable or an astable multivibrator. Although it is not in the category of TTL devices, yet it can be used with other TTL devices. It operates at voltage ranging from 5 V to 15 V, which indicates its wide adaptability.

The IC is composed of two comparators—a flip-flop, an output stage and a reset circuit. The figure below shows the IC wired as an astable multivibrator. The output frequency depends upon the values of R2 and C, while the duty ratio is determined by R1and R2. The internal structure of this popular device is given in Figure 5.18.

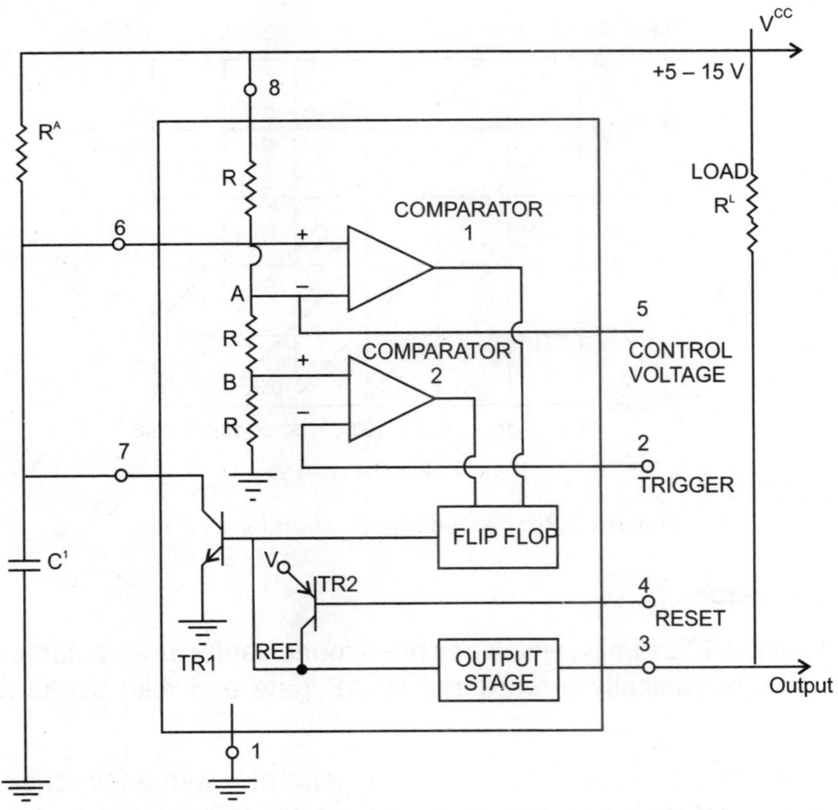

Figure 5.18 *Internal structure of IC 555*

This IC is also widely used as a timer, providing a wide range of timing possibilities. A clock pulse generator based on IC 555 is shown in Figure 5.19. This circuit is capable of generating very low frequencies—0.1 Hz to a frequency of 1 MHz by a suitable choice of resistors and capacitors.

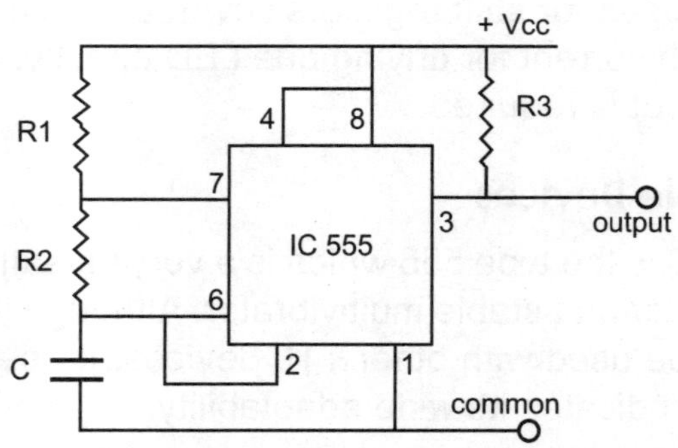

Figure 5.19 *IC 555 wired as astable multivibrator*

By varying R1 and R2, various duty ratios in the output signals may be obtained. A frequency/RC chart is given in Figure 5.20.

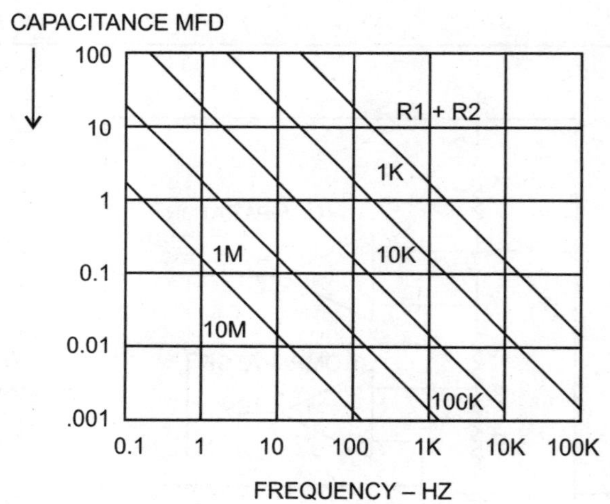

Figure 5.20 *Frequency/RC chart for IC 555*

Astable Multivibrator

One type of IC of the TTL family, which has been found useful as an astable multivibrator is the 7413. This is basically a four-input NAND gate and also useful as a Schmitt trigger.

Figure 5.21 shows this IC connected as a clock generator with a feedback resistor and a capacitor, which determine its operating frequency—the output being in the form of rectangular pulses. With different values of C, the frequency can be varied over wide limits. In practice a frequency range of 1 Hz to 1 MHz is easily attainable. The 7413 package consists of two gates. Each of these can be used independently.

7413 based pulse generator: The oscillator is simple and utilises one half of the IC, together with one resistor and one capacitor.

The resistor value is fixed at around 390 ohms. The value of the capacitor determines the clock frequency, typical values for which are as follow s:

1000 Mfd........2 Hz	1 Mfd.............2 KHz
100 Mfd..........20 Hz	0.1 Mfd...........20 KHz
10 Mfd...........200 Hz	.01 Mfd..........200 KHz

The fan out (output loading factor) is 10.

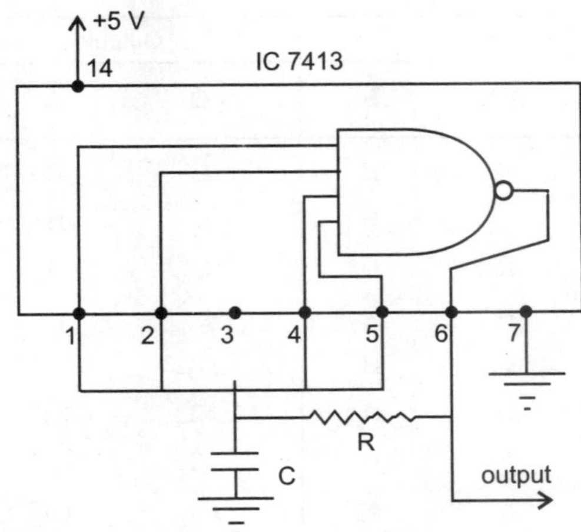

Figure 5.21 *IC 7413 as a clock pulse generator*

If a number of capacitors are switched through a rotary switch, a wide range of frequencies can be generated. For reliable operation, long leads should be avoided in connecting capacitors used for frequency selection. The capacitors should be mounted as close to the IC as possible.

Since 7413 package has two independent gates, two separate oscillators can be built around one IC.

Monostable Circuits

IC type 74121, a single pulse generator, has two triggering modes:

(a) a positive trigger at pin 5 and

(b) a negative trigger at pins 3 and 4

This is shown in Figure 5.22 below along with the function table, which indicates the behaviour of the IC in different conditions.

H and L are high and low voltage levels, while X is 'don't care'. The arrows indicate the input pulse triggers, low to high and high to low transitions.

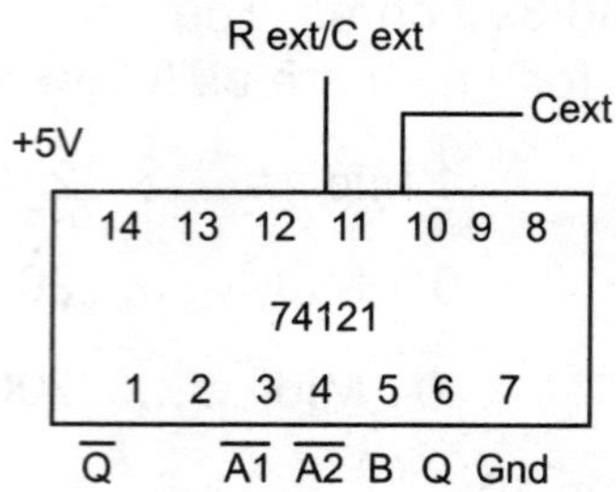

Function Table

Input			↑ Outputs	
$\overline{A1}$	$\overline{A2}$	B	$\overline{Q}$	$\overline{Q}$
L	X	H	L	H
X	L	H	L	H
X	X	L	L	H
H	H	X	L	H
H	↓	H	⊓	⊔
↓	H	H	⊓	⊔
↓	↓	H	⊓	⊔
L	X	↑	⊓	⊔
X	L	↑	⊓	⊔

Figure 5.22 *Monostable IC 74121 with function table*

Once triggered, the output pulse width depends on the RC values. If the value of the resistor R exceeds a certain limit—approximately 150K, then triggering becomes unreliable.

Two outputs are available from this device—a positive mono-pulse is available from 'Q' output, while an identical negative pulse is available from the Q^{-}. The two outputs may be used simultaneously to drive other circuits. Output pulse length may be varied from 20 nanoseconds to 28 seconds, by appropriate choice of timing components.

Short duration relay operation is possible if the pulse width is long and if a transistorized relay driver is interfaced.

Monostable Multivibrators (Type 74121)

This IC can be used as a monostable circuit, capable of being triggered by both negative and positive going pulses. Two pulse outputs are provided—a positive-going pulse from the 'Q' output and a negative-going pulse from 'Q̄' output. It has an added advantage that output pulse widths can be varied over a wide range, when appropriate RC elements are used externally. This IC can also generate precise pulse widths for calibration purposes.

Phase-locked Loops

The phase-locked loop (PLL) contains an electronically-controlled oscillator, which can be synchronized with an incoming signal. The internal structure of the PLL IC Type 567 is shown in Figure 5.23.

When an input signal is applied to the IC, the phase detector will compare phase of the input signal with that of the signal generated by the current controlled oscillator.

The error voltage at the output of the phase detector is dependent on the phase difference between the two signals. It may either be positive or negative, according to which of the signals is leading or lagging the other in phase.

The error signal is passed through a low pass filter, after which, it is amplified to produce the signal which is used to control the frequency of the oscillator. The control signal alters frequency of the current controlled oscillator and can make it approach the frequency of the input signal. If the input frequency is close enough to the free-running frequency of the current-controlled oscillator, the latter will become locked to the input frequency and the output goes low.

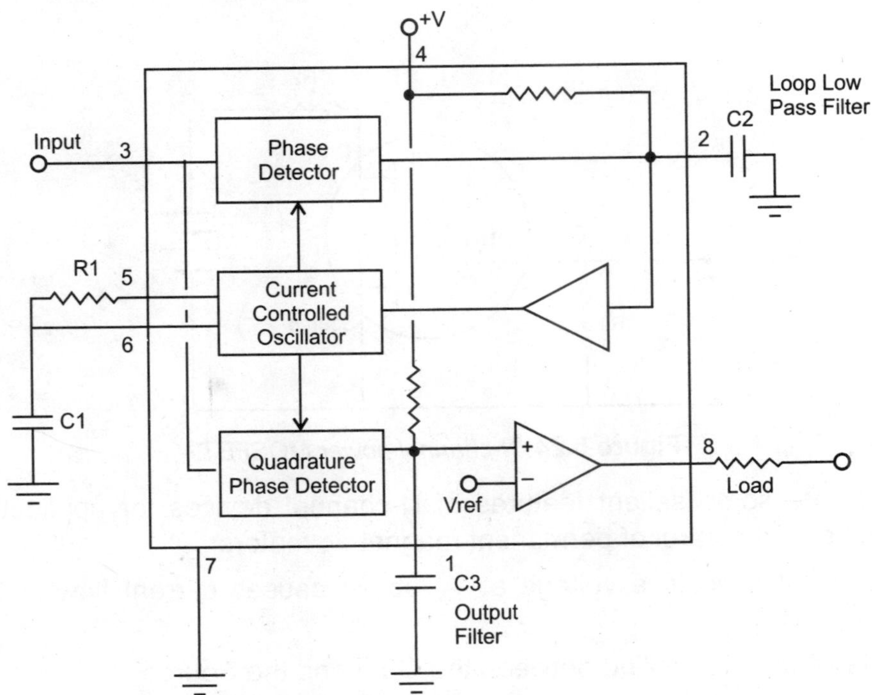

Figure 5.23 *Internal structure of 8 pin PLL IC type 567*

Power MOSFETS

A MOSFET is a Metal Oxide Semiconductor Field Effect Transistor. Power MOSFETS are later generation semiconductors and well suited to drive DC motors. However, some precautions are necessary in the design of their circuits and in practical usage, otherwise experiments which result in damaged devices will prove costly.

Power MOSFETS are several times more costly than transistors of comparable performance.

Unlike transistors that are current operated devices, power MOSFETS are voltage operated—similar to the older generation vacuum tubes.

Two types of power MOSFETS are the N-Channel and the P-Channel type; the N-channel being more common and having a wide range of applications.

Power MOSFETS are three electrode devices—the 'Drain', the 'Source' and the 'Gate'. In N-channel devices, a positive voltage is applied to the 'drain' while the 'source' is at the negative end. In absence of any voltage on the gate, the device is cut off. On the application of a positive voltage to this gate, the device conducts and current flows in the drain-source circuit. A minimum voltage is required to set the MOSFET in conduction. This is called the threshold voltage. A voltage higher than this is required for full saturation and the device acts like a switch with very low resistance. It is due to this switching property that dc motors can operate in pulse width switching (PWM) mode.

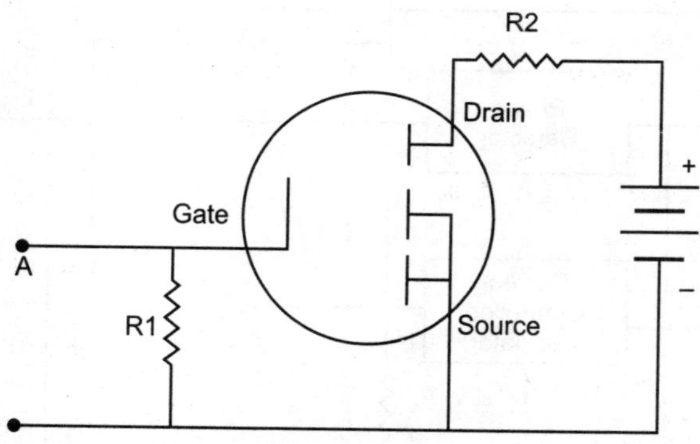

Figure 5.24 *N-channel power MOSFET*

Listed below are some salient features of N-channel devices for application in the direction and speed control of permanent magnet dc motors:

(a) Application of a positive voltage at 'A' above causes current flow in the Drain-Source circuit.

(b) The gate voltage is applied between the Gate and the Source.

(c) Application of a positive voltage at 'A' above causes current flow in the Drain-Source circuit.

(d) The gate voltage is applied between the Gate and the Source.

(e) The Gate current is negligible (depending on R1). However, R1 may be eliminated.

(f) A minimum gate voltage is required to start the conduction process. This is known as the threshold voltage.

(g) For saturation (full conduction), a slightly higher gate voltage is required.

(h) Beyond a specific voltage the Gate breaks down and the device is damaged.

(i) The Drain-Source voltage must also be kept within specified limits.

(j) The Drain current is limited - on saturation–by the battery voltage, the resistance R2 and the Drain-Source resistance. This current must also be maintained within a specific value for the Drain-Source voltage.

(k) For driving inductive loads on pulse drives, a snubber must be used to eliminate negative spikes which are generated due to current cut-off on the falling edge of the pulse.

(l) Performance curves of the MOSFETS are available for all devices (from the manufacturers). These must be taken into account while designing the circuits.

(m) In pulse circuits, the devices run cool as no heat is generated. However, for heavier currents, adequate heat sinks and proper ventilation for the devices should be provided.

6 Optoelectronic Devices

Light-emitting Diodes

The most widely used optoelectronic devices are the light-emitting diodes or LEDs. These are semiconductor devices and when suitably forward biased, emit light of different colours, depending on the specific LED in use. Common colours are red, green, yellow, amber and blue. Combinations of different colours in a single package yield a light of white colour. Infra red LEDs are also available. These find extensive applications in remote controls for television and other devices.

Compared to other sources of light, LEDs have some distinct advantages. They operate cool and can be used for more than 10,000 hours. Their current consumption is low and they have fast switching response—ideal for digital circuit applications. Although the earlier LEDs had a low light output, new developments have offset this disadvantage to a considerable extent. For instrumentation and interior displays, they are unparalleled.

LEDs come in various sizes, from the smallest having a diameter of 1 mm to 10 mm and growing larger as new developments take place. While most LEDs have a circular cross-section, LEDs with rectangular cross sections are also available. When these devices need to be closely spaced, rectangular ones are to be preferred. Since they emit only one colour, combinations of these in a single package result in multi-coloured emissions. Such combinations have LEDs which are connected in parallel, with either anodes or cathodes connected together.

When LEDs of different colours are wired in parallel, they should have the same operating voltage, otherwise the LED with a higher operating voltage will not function due to the supply voltage being 'clamped' at a lower level. One way to avoid this is to use separate resistors in the anode circuit, as shown in Figures 6.1 and 6.2.

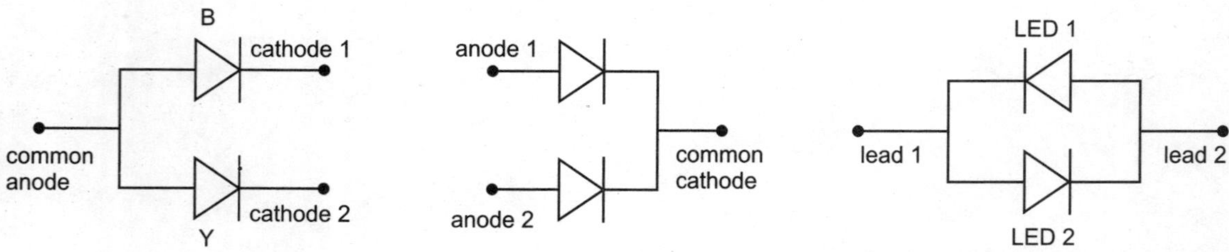

Figure 6.1 *Electrical connections in multiple LED packages*

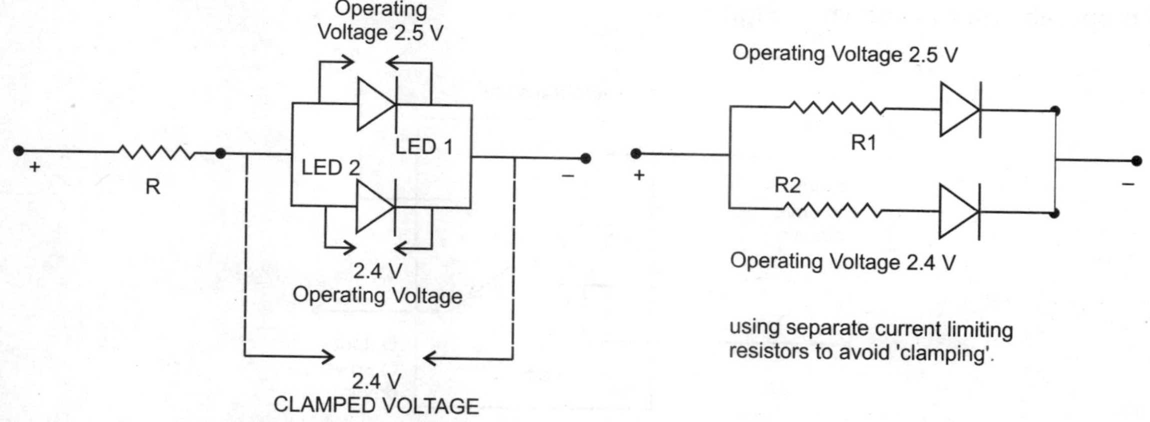

Figure 6.2 *Paralleled LEDS—Clamping and a solution*

The active semiconductor wafers which produce the light are encapsulated in a plastic material which is very sensitive to high temperatures. If the lead wires from the encapsulation are overheated while soldering, the plastic covering melts and the LED is rendered useless.

The low operating voltage of LEDs renders them safe to use in hazardous environments. When displays comprising several LEDs are fabricated, multiplexing techniques are useful since they reduce the number of ICs in the overall circuit.

Opto-isolators

We now introduce an opto-electronic device which is known as an opto-isolator. When logic circuits have to interface with other high power devices like power MOSFETs, at times problems may arise as they cannot be connected directly. DC motors are driven either through power semiconductors or through relays which are driven by relay drive circuits. Electromagnets and DC motors generate electrical disturbances which travel through the voltage supply lines and reach the logic circuits. Known as glitches, these noise signals are of very short duration. They are generated during the 'switch off' instant of the electromagnetic devices as the current changes peak at such times.

A device called opto-isolator will solve such problems. Opto-isolators have two active components in a single light-tight package—LED that connects to the logic drive circuit and phototransistor that receives the optical signal from the LED and connects to the power drive circuit. Since the two components are connected to different power supply lines, electrical isolation is achieved (the signal transfer taking place via the optical route).

An opto-isolator is shown in Figure 6.3.

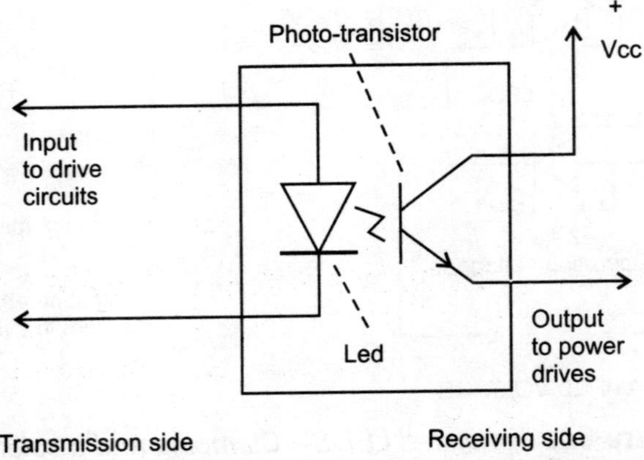

Figure 6.3 *Single channel opto-isolator*

The power supply lines have to be isolated and a common ground is to be avoided. This enables us to electrically isolate the input and the output circuits.

7

Fibre Optics

Thin, transparent plastic strands constitute what are known as optical fibres; the relevant technology is called fibre optics. They appear like webs spun by spiders and are manufactures in various thicknesses, ranging from 10–100 μm; the 15 μm diameter being the most commonly used. Diameters more than 50 μm are stiffer and cannot be bent or twisted readily. Light passes through these fibres by the means of multiple total internal reflections, the reflections taking place at the inside of the fibre walls. See this Figure 7.1 in coloured illustrations on pages 177–184. The efficiency of light transmission is very high and only a small fraction of the light intensity is absorbed during the reflection process.

Optical fibres offer special advantages. Since they are synthetically fabricated, costly electrical transmission metals like copper are avoided. The signals are optical in nature; hence hazards commonly encountered with metal wires are not present.

The frequency range of these fibres is from the visible to the near infra red, while the operating temperatures are from –40 °C to +70 °C. Uncased fibres can be bent from 12.5mm radius onwards.

In special projects, fibres can be used in front displays while the illuminating LEDS can be tucked away inside the enclosure at any convenient place.

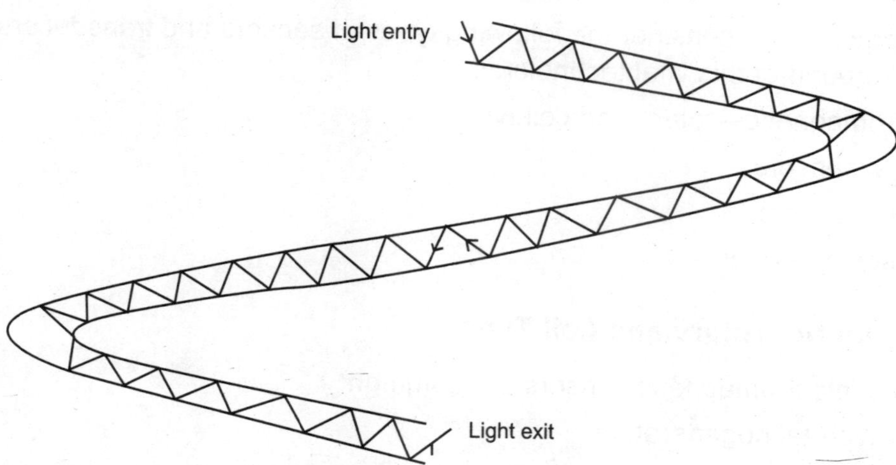

Figure 7.1 *Light travelling by total internal reflections in an optical fibre*

8 Sensors and Transducers

Sensors are devices that 'sense'. They detect signals of one type and convert these into signals of another type—mostly electrical.

Presently, there are several types of sensors, covering almost all areas of technology and new types are also appearing regularly. They range from the simplest, like those used in remote control of TVs, to the very advanced used in airplanes, space ships, automobiles medical and nuclear instrumentation, etc.

Although the terms sensors and transducers are used synonymously, yet this is not strictly correct. To give a simple example, consider the TV remote control system. It consists of two parts—the hand held remote controller produces an IR signal (coded), which is picked up by a phototransistor in the TV and converts this into the relevant electrical signal to control the various functions of the TV set.

In this example, the device which produces a non-electrical output from an electrical signal in the hand-held remote controller is a transducer. On the receiving end, the device which senses the infra-red from the remote controller and converts this into an electrical signal is a sensor.

In some cases, there are devices which serve both as sensors and transducers; ultrasonic crystals are one such type.

In this chapter, we will consider the following types of sensors and transducers so as to give a background of this vital technology:

1. Electromagnetic—rotary and coil type
2. Optical sensors
3. Tactile sensors
4. Ultrasonic devices.

Electromagnetic—rotary and Coil Type

Two types of electromagnetic sensors are common:

(a) Rotary type tachogenerators

(b) Stationary coil type induction sensors

Rotary type Tachogenerators

These are basically dynamos. They generate a voltage—DC or AC, depending on the type, which is proportional to the rate at which they are revolving. Their basic application is for monitoring the RPM and to use this information for controlling drive motors. As such, they are speed sensors. However, they are not expected to deliver usable power, unlike conventional dynamos. Tachos are of two types:

i. Integral—those which form an integral assembly with the motors
ii. Stand alone—those which have to be mechanically coupled to the motors/drive mechanisms for monitoring the rpm/speed of mobile robots.

In the former case, tachogenerators will give an output voltage proportional—generally linear—to the motor's RPM, irrespective of the integral gearbox (if any). Where stand-alone tachos are used, they will monitor the RPM of the mechanically connected devices, such as gear box output shaft, drive wheels, etc.

The following specifications are generally available from the manufacturer's data sheets to enable the design of tacho based speed monitoring/control systems:

(a) Linearity—Linearity of output voltage v/s RPM.
(b) Armature resistance—DC resistance of armature winding of the tacho in ohms.
(c) Armature inductance—Inductance of armature winding in mH.

Following points are to be noted in connection with these devices:

i. A permanent magnet DC motor reverses direction if the polarity of voltage applied to the motor at its terminals is reversed. In a tacho, if the direction of rotation reverses, the polarity of voltage at its output also reverses. By this means, the direction of rotation of the motors can be sensed.
ii. Tachogenerators use metallic brushes alloyed with silver to reduce the voltage drop across them.
iii. V-out is proportional to the RPM. Even at low speeds, a voltage is generated, which can be used for control applications. A standard tachogenerator is designed to give an output of 7 V per 1000 RPM.

The figure given below indicates one possible method of speed control using tachogenerators.

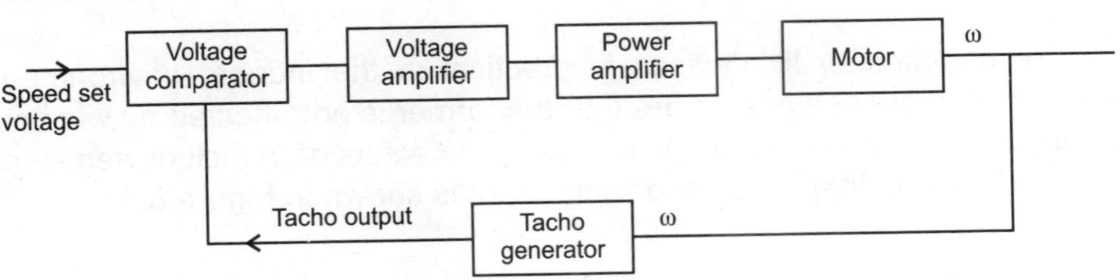

Figure 8.1 *An arrangement for speed control using a tacho*

Stationary Coil type Induction Sensors

An induction sensor is a simple tuned circuit consisting of a coil and capacitor arrangement (as shown in Figure 8.2).

This simple device serves as a sensor for electrical oscillations in its vicinity. The frequency of oscillations picked up will depend on the respective values of L and C, and can be calculated from standard formula. For frequencies in AF range, the coil should have a high inductance value, by using a ferrite core.

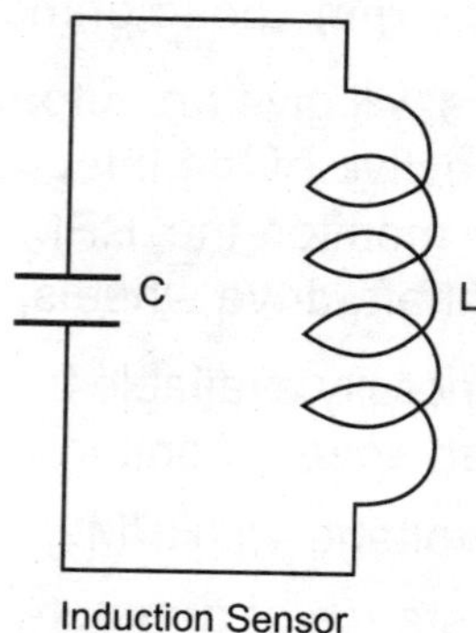

Figure 8.2 *Inductor (L) and capacitor (C) form an induction sensor*

Two such sensors in a proper geometrical arrangement serve to guide vehicles over pre-laid tracks. These automatic guided vehicles are of importance in factory automation systems.

Optical Sensors

Infra-red (IR) radiation is invisible and lies beyond the red end of the visible light spectrum. This type of radiation is emitted by heated substances and warm-blooded animals.

Artificial emitters of IR include surgical lamps, IR LEDs, and IR lasers. For remote control purposes, IR LEDs are quite suitable and a wide range of control mechanisms are possible. In the domestic environment, remote control systems using IR devices are used for controlling functions of a variety of entertainment systems, garage door openers, etc.

The range of commercial IR devices is practical in the indoors environment. The advantages of IR over US systems are that the former is not affected by wind currents and that control is possible through glass panels. Detectors of IR include semiconductor photo diodes. An IR emitter/receiver arrangement is shown in Figure 8.3.

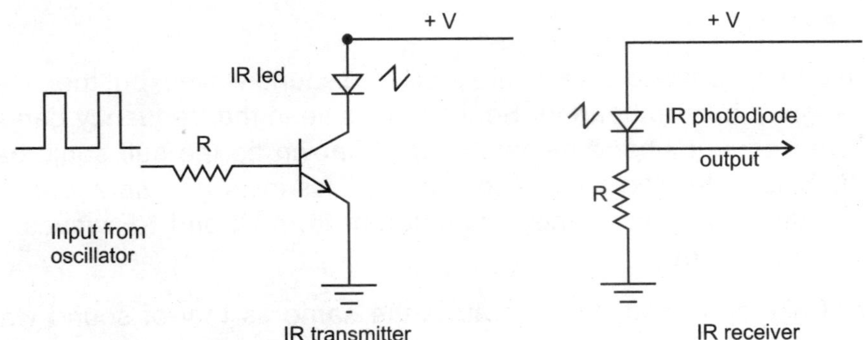

Figure 8.3 *Optical transducer-sensor pair*

Tactile Sensors

Tactile or touch sensors constitute an integral part of most technologies. Such sensors may be simple limit switches, conductive foam pads, sensing whiskers, etc.; depending on what parameter you want to sense.

Tactile sensors are different from other types of sensor in which the sensing function is operative only on contact. Thus, a force sensor is a tactile sensor; a contact thermometer is a tactile sensor; and a gramophone needle is a tactile sensor.

A human being has five senses—vision, hearing, taste, smell and touch. Taste also has to do with touch, since any substance which is identifiable by taste has to make contact with the tongue. It is different from tactile properties of the skin, since the taste buds are located only on the tongue.

Tactile capability in humans is quite advanced. The fingers can detect any solid substance (contour sensing), distinguish between applied forces (force sensing), distinguish between hot and cold objects (temperature sensing) and feel minute vibrations (vibration sensing). These senses along with the sense of hearing become more acute in people suffering from visual handicaps.

In robots, tactile sensing is of importance in grippers. The gripper, when required, should be able to crack a nut and also to handle an egg without breaking it. Where tactile mapping is required, these sensors enable the contours of a solid object to be determined.

Slip sensors, another form of tactile sensors, enable a gripper to adjust its gripping force if it detects that the object being held is slipping down.

Elementary tactile sensors, in form of limit switches, prevent over-riding of arms and manipulators, so that they do not damage the joints and linkages. In mobile robots and automatic guided vehicles, such tactile sensors in form of bumpers, stop further movement when an obstruction is detected. Thus, tactile sensors range from very simple devices like limit switches to quite sophisticated ones. Tactile sensing technology does not consist of just the sensors. They may also need electronic processing circuits and in some cases microprocessors with appropriate software.

Ultrasonic Devices

Ultrasonics, meaning 'beyond sound', are similar to sound waves, but their frequency is beyond the range of human hearing. Sound waves lie in the frequency band of 16 Hz to 16 KHz. The frequency band below 16 Hz is known as the sub-sonic band, while that above 16 KHz is termed ultrasonic. Since ultrasonics are vibrations in material media (solids, liquids or gases), they are different from IR and RF waves, which are electromagnetic radiations.

The velocity of ultrasonic radiations in air is the same as that of sound waves—340 meters/second. In this respect, they travel slower than IR or RF radiations which travel at the velocity of light—300 km/sec. In remote control and obstruction detection applications, this is not a limitation since the distances over which the controls are exercised are very limited.

Ultrasonic sensors and transducers are basically of two types:

1. Piezo-electric devices
2. Magnetostriction devices

For control and sensing purposes, where large power is not required, the piezo variety is commonly used. Naturally occurring quartz crystals, when properly cut and finished, exhibit the piezo-electric effect. For generation of ultrasonics, quartz is not as efficient as the synthetic barium titanate (BaTi) or the lead-zirconate titanate (PZT). The more common ultrasonic transducers are of the latter variety. The transducer works as an emitter as well as a receiver. Occasionally, the same crystal is used for both these functions.

Every crystal works in a narrow frequency band—the resonant frequency. When an electrical signal of this frequency is applied to the crystal, it oscillates at an identical frequency producing the mechanical ultrasonic waves. Conversely, when ultrasonic waves of the resonant frequency impinge on the crystal, electrical signals of this frequency are produced across the crystal.

For use in air, the frequency that is found most suitable is 40 KHz, and most ultrasonic operations in this medium use 40 KHz crystals—either as a matched pair for transmission and reception or as a single crystal performing both functions when used in the pulsed mode. For obstruction detection in mobile robots, the property of reflectivity of sound is utilised. See Figure 8.4.

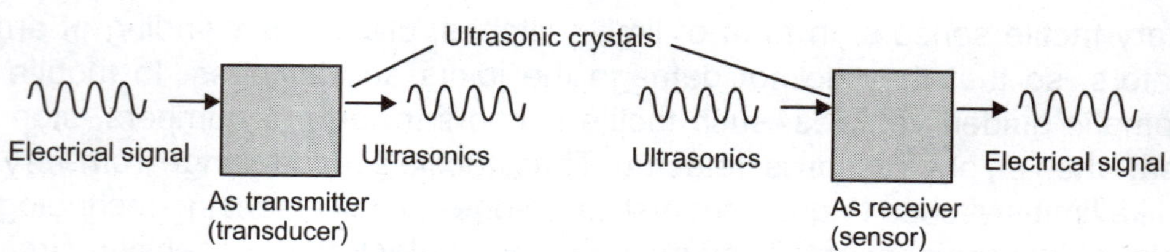

Figure 8.4 *Behaviour of ultrasonic crystals*

9

Special Alloys—Muscle Wires

We are familiar with devices which produce motion of one kind or other through electrical energy. Examples are electromagnetic solenoids which produce linear movement and electric motors which produce rotary motion. While their actions are different, they have one feature in common–they use electromagnetism to produce the required result. In the former case, a coil of wire wound on an iron core produces the required effect. In the latter case, it is an armature wound with a coil of wire and located in a magnetic field which does the work. These devices have been around for more than a century and are to be found in almost all domestic and industrial appliances.

Recently, new ways to create motion using electric energy have been developed. This is a totally different approach and uses special alloys – mixtures of metals to perform this task. These alloys change their shape when heated or cooled and exert a tremendous force in this process.

While shape memory alloys come in different forms, we shall be dealing only with a special type—an alloy in wire form known as muscle wire, which is of particular interest in robotics applications. When heated, these wires contract in length, exerting a tremendous force which can be used to perform a wide variety of operations. They are capable of lifting weights, which are thousands of times their own weight. This gives us the potentiality of these devices. Another feature of these wires is that they can operate without emitting any sound, unlike the electromagnetic solenoids or the electric motors. Further, they contract very fast and with precision—an important property when they are incorporated in the design and fabrication of systems that should operate with exactness.

We shall present here a simple mechanism that demonstrates the basic action of the muscle wires.

In the figure below, a length of muscle wire is shown suspended from a hook. At the lower end of the muscle wire a weight is hung, which under action of gravity keeps the muscle wire taut.

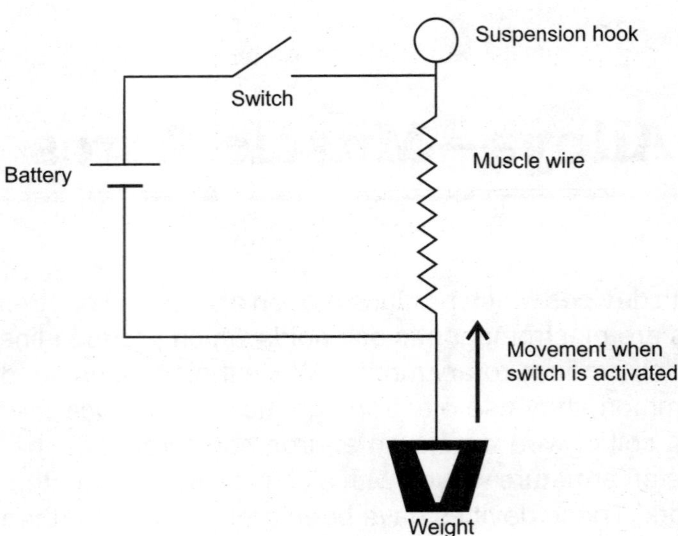

Figure 9.1 *Simple demo of muscle wire action*

When current from the battery is switched on, the wire contracts and pulls the weight upwards. When the current is switched off, the wire regains its original length and the weight returns to the position where it was, before the current was switched on.

The vertical movement of the weight; the minimum current required to activate the muscle wire; the maximum safe current which the muscle wire can handle; all these depend on the diameter and length of the muscle wire being used. For example, a muscle wire of diameter of 100 μm and 10 cm length can lift a weight of 150 grams in 0.5 seconds, when electrically operated. The power consumption then is only 0.5 watt. Ten of these wires operated in parallel will lift 1.5 kg during the same duration, consuming 5 watts of electrical energy.

On the other hand, if we use 250 muscle wires of 250 μm diameter each, the total lift will be approximately 232 kg in 0.5 seconds with a power consumption of 500 watts!

It is this property of muscle wires—the snap action and precise contraction when activated—which makes the application of muscle wires so important in robotics.

Although muscle wires are very conveniently activated by passing an electric current through them, they can also be activated by an external heat source, like an open flame, solar energy, etc. In any case the maximum safe temperature specified by the manufacturer should not be exceeded; otherwise, the muscle wire would be permanently damaged.

As we have mentioned earlier, muscle wires belong to a class of alloys called shape memory alloys, which possess a crystal structure that is altered at a specific temperature, termed the transition temperature. Below this temperature, these alloys can be stretched without damage. On heating above this transition temperature, the alloy returns to its original condition causing the wire to contract. This contraction on heating the wire is about 8% of the length of wire used.

Shape memory alloys have been made using combinations of different metals—the most common alloy from which muscle wires are fabricated is nitinol, composed of nickel and titanium metals.

Although we have demonstrated the 'pulling' action of muscle wires in the above example, yet levers can be fabricated for a host of applications in robotics. By suitable arrangements, the movement of muscle wires can be magnified several times using appropriate mechanisms. By themselves, the muscle wires operate silently, unlike electric motors and solenoids. They do not generate any electrical disturbances due to sparking at the armature brushes.

Even though shape memory alloys are available in different forms, such as rods and sheets, these are not convenient to work with due to their hardness. Hence, alloys in wire form are most commonly used by experimenters who have limited facilities at their disposal. The wires can be cut easily and bent (within limits). When used within safe temperature limits, these wires have a long life and can be used repeatedly hundreds of times.

Properties of Muscle Wires

Muscle wires are thin strands of nitinol—an alloy of nickel and titanium metals. They come in different sizes (seven at present) and two transition temperatures - low temperature and high temperature. Transition temperature is the point where the alloy changes from its easily deformed low temperature phase to its harder high temperature phase. The low transition temperature is at 70 °C, while the high transition temperature is at 90 °C. The wires can be easily stretched at room temperatures by using a small force. When heated, the wire changes to a much harder form and contracts with a substantial force. It is this property of contraction on heating that is utilised in development of various devices in robotics and other areas.

An important parameter in the use of muscle wires is the amount of contraction that these wires undergo, when heated. They can be stretched up to 8 per cent of their length with full recovery, but limited period of operation. If this contraction is restricted to 3 per cent (up to a maximum of 5 per cent), these muscle wires can easily operate for millions of cycles, without interruption.

The larger the wire size, the stronger is the wire. Two parameters here are the recovery force and deformation force.

Let us consider two wire sizes and compare their properties, which will provide some idea of how these depend on the wire diameters.

Size: Diameter in µm:	100	300
Maximum Recovery Force in grams:	470	4,240
Recommended Recovery Force in grams:	150	1,250
Recommended Deformation Force: grams:	28	245

What do these figures mean?

Maximum Recovery Force (MRF)

MRF specifies the largest force that the muscle wire can exert when it is heated. However, this figure is not what is recommended in actual operation. To protect the wires from overstrain and maintain their characteristics for repeated operations over an extended period of time, only a third of this value is recommended.

This is evident from the figures given for wire size of 100 µm, where the MRF is given as 470 g and the Recommended Recovery Force (RRF) is 150 g. For wire size of 300 µm the MRF and the RRF are 4,240 and 1250 respectively.

But what is the Deformation Force?

Also known as the Bias Force, this is the force that should be applied to the wire after it has cooled, to restore it to its original length. In absence of this force, the wire will not return to its pre-heated state.

In using the muscle wires for performing actions, they may require to be bent one or more times in a final physical configuration. In such cases, the wires should not be bent sharply to avoid overstraining the wires. For the two examples considered above, the minimum bend radius for the 100 µm wire is 5 mm while for the 300 µm wire it is 15 mm. The bend radius is clarified in the following figure:

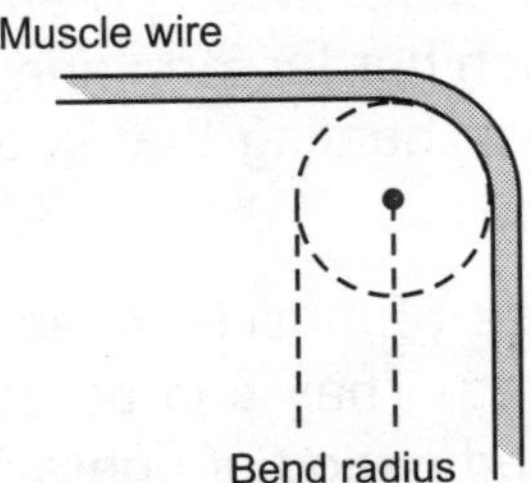

Figure 9.2 *Bend radius of muscle wires*

We will now consider some electrical properties of the muscle wires. The linear resistance is specified in ohms/meter length.

Wire size	100 µm	300 µm
Electrical resistance per meter	150 Ω	13 Ω
Recommended current	180 mA	1,750 mA

How fast do these wires contract? The following figures illustrate this.

Low temperature	100 μm	300 μm
Contraction speed	0.1 sec	0.1 sec
Relaxation speed	1.7 sec	8.5 sec
Thermal cycle rate	33 cyc/min	7.0 cyc/min

From the above given data provided by the manufacturers in their data sheet, we draw the following conclusion:

From the point of thermal cycle rate, the relaxation speed and minimum bend radius, it is best to use thinner wires—several of them in parallel, wherever possible. In actual practice, we must ensure that mechanical and electrical connections are proper and that during soldering operations, heat sinks are used to prevent the wires from overheating and getting damaged in the process.

10 Remote Control Techniques

Remote control between operator and an object implies control without a physical link. We shall consider two techniques for remote control of various devices. These are:

(a) Infra-red control techniques

(b) Radio control systems

Infra Red Techniques

A remote control system consists of two units—a transmitter and a receiver. The transmitter may be hand held; when pointed towards the receiver, will send a beam of pulsed IR radiation which will be intercepted by the IR photo diode on the receiver unit. This is one of the disadvantages of using IR to control devices—the control is possible only in line-of sight operation.

The transmitted pulse must have stable frequency. If it is coded using pulse code modulation techniques, then several operations are possible with one hand-held device. The circuit shown below is relatively simple. It generates a single frequency with the components used.

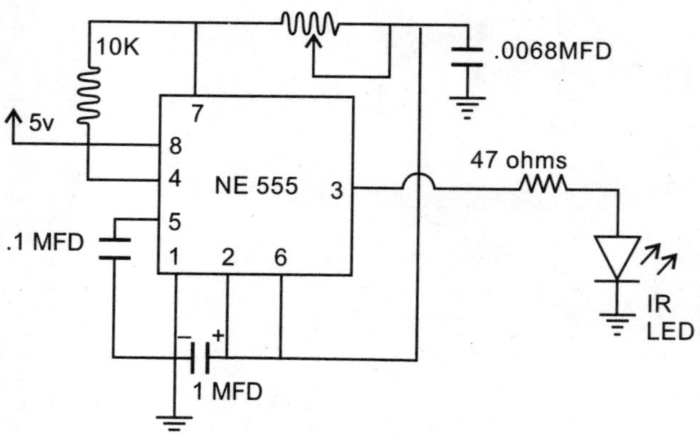

Figure 10.1 *IR transmitter*

With the circuit shown, the frequency can be altered within the given range by varying the 100k variable resister. In case more than one frequency is desired, then the variable resister should be replaced with fixed resisters and switches introduced to select different values.

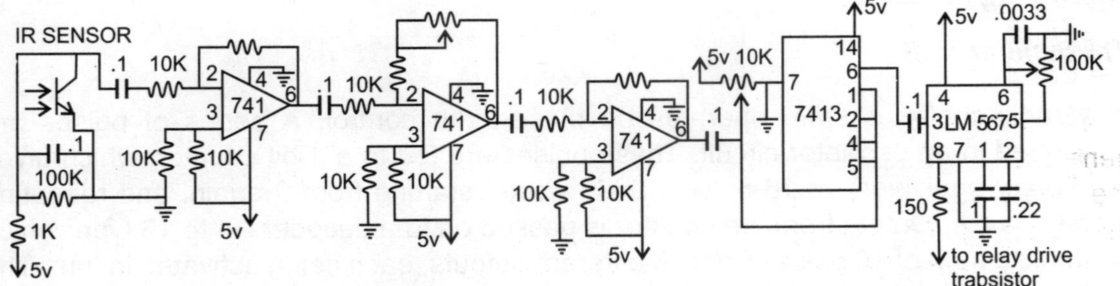

Figure 10.2 *IR receiver for remote control applications*

The IR beam, when directed at the receiver, will be intercepted by a photodiode in the receiver circuit shown above.

This signal is amplified by three stages of amplifiers and operates a Schmitt trigger (7413) to produce pulses of equal amplitude, which are then fed to the tone decoder IC 567. This has to be tuned, through 100k variable resister, to the frequency of the incoming IR beam. When the frequencies match, the output of the 567 will go high. This can be used to operate the motor control circuits shown earlier, or some other device, as required. For multi-channel operation, when different frequencies are used for control purposes, more stages of the 567 tone decoder need to be added in parallel, each output of which performs a different function.

Radio Control Techniques

In radio-control systems, some techniques described earlier can be used. Depending on the number of control channels required, a coder has to be designed which generates the necessary control code. This coded signal is modulated with an RF carrier and transmitted. On the receiver side, the coded signal is extracted from the modulated RF carrier and is used to operate the various devices which control the robot functions, e.g., relays, potentiometers, etc. The relays, in turn, switch on and switch off the motors, and reverse their direction, while the potentiometers can control the speed of the motors. One of the advantages of radio-controlled system is that line-of sight operation is not required. Yet another advantage is that it can operate over very large distances.

In Figure 10.3 below, a radio-control system of the type used for controlling the movement and other operations in a mobile robot, is given.

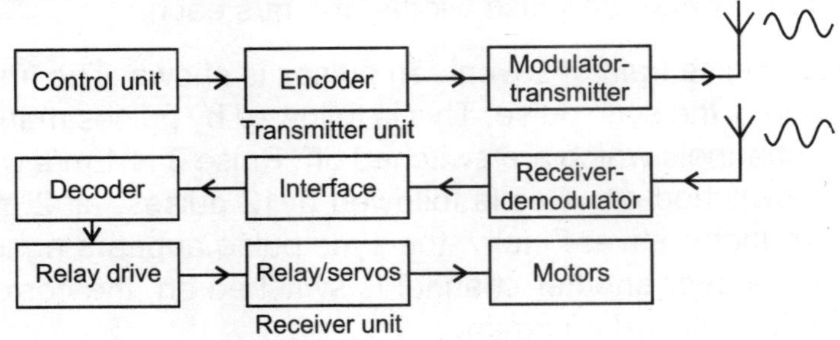

Figure 10.3 *Radio control system*

Operating Principle

Transmitter Unit

T e radio-control system operates on pulse-width control. A series of pulses are generated in an oscillator circuit. These pulses are fed to a 4-bit counter, which gives the binary equivalent of 16 pulses; then resets, starting from 0 again, and repeating the sequence. Output from the counter is passed on to a decoder—4 to 16 type, which provides a train of 16 pulses from 16 different outputs, each being activated in turn. The first output provides a synchronous pulse, which synchronises the receiver circuit also. This is followed by 15 pulses whose width can be variable, depending on the control desired. These pulses are modulated with a carrier wave in the 27 MHz experimental band and transmitted.

The encoded pulses carry the information required to perform the control operation, which is dependent on the pulse width of each of the 15 encoded pulses. See Figure 10.3 below.

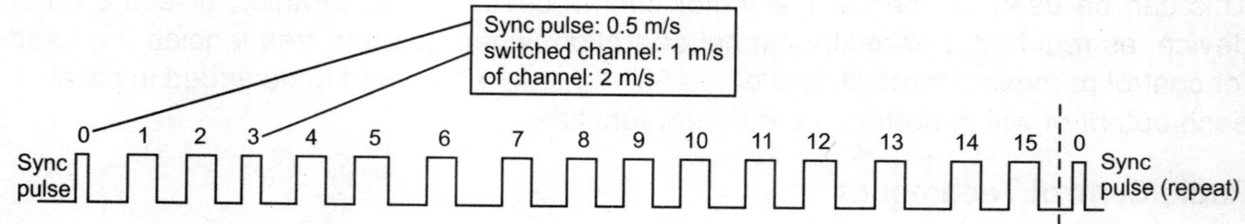

Figure 10.4 *Encoded RF (Radio Frequency) pulses for radio control*

This is the time-division multiplex principle; the significance of the pulse widths is as follows:

(a) When the receiver-decoder receives a pulse of 0.5 m/s width sync pulse, the counter is reset to '0'. The next pulse is the first channel, the one after this the second channel till the last pulse in this train—corresponding to the15th channel is received, after which, the sync pulse again appears and the pulse train recommences, starting from the first channel again.

(b) Channels which are not operative generate pulse widths of 2.3 m/s each. Channels which are operative generate pulse widths of 1 m/s each.

In the above figure, a pulse train of seventeen pulses is shown. The first pulse, marked '0' is 0.5 m/s wide and is the sync pulse. This is followed by pulses marked 1, 2, each of 2.3 m/s. These are channels which are switched off. Pulse 3 is 1 m/s wide and denotes that this channel is switched on. This is followed by12 pulses, all 2 m/s wide; hence, channels 4 to 15 are inoperative. Finally, the sync pulse appears again and the pulse train is repeated. Whenever another channel is switched on, the corresponding pulse width changes, till it is switched off again.

Although we have shown 15 operating channels here, less or more channels could be provided by suitably altering the circuits—specifically the counter-decoder pair of ICs.

A pulse coder circuit is shown in Figure 10.5 below, which uses standard ICs. This circuit generates the pulse train for control purposes and provides 15 channels of operation. We will present only switched channels here. This enables us to control 15 relays in any device to perform the various functions—forward and reverse operation of the motors.

The coder is a vital part of the radio control system, since it generates the pulses which carry all the information for control purposes. The circuit operation is as follows:

Three transistors form a stable multivibrator and act as a pulse generator. The base of the second transistor is also connected to the common output of the 4–16 channel decoder, IC type 74154. The pulse width of each channel is set by the 0.22 Mfdcapacitor and the corresponding resistance of that channel on the decoder IC. This is elaborated below:

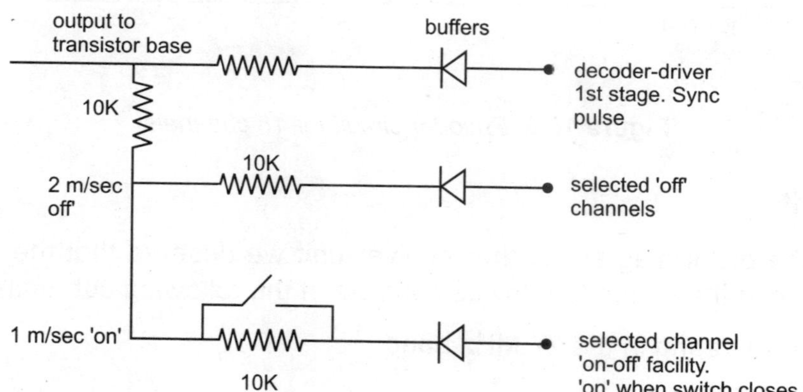

Figure 10.5 *Coder operation*

It may be seen that in any channel, if a switched (on-off) control is required, a toggle switch is wired across the appropriate 10k resistor. If the channel is not required, the switch may be omitted.

The output pulse train is from the common collector of the transistors and inverted twice using two NAND gates of the 7400 IC. This yields clean pulses which are fed to the modulator section of a low power, 27 MHz RF transmitter.

Several transmitter circuits are described in electronic journals; all of them cannot be covered here. However, when selecting an appropriate circuit, it should be thoroughly checked that provisions for modulation are available so as to mix the coded signal with the carrier wave. See Figure 10.6 given below.

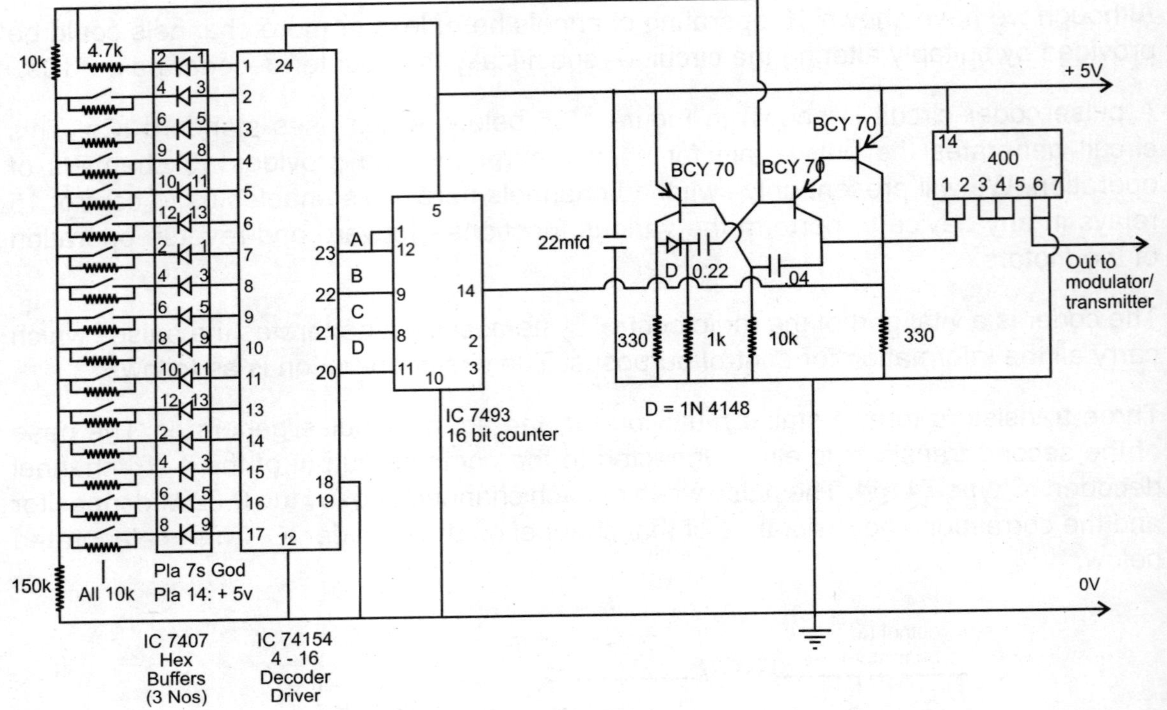

Figure 10.6 *Encoder circuit for 15 channels*

Receiver Unit

Referring to the block diagram of the receiver unit we observe that the radio control receiver system in the controlled device consists of the following sub-units:

(a) A receiver operating in the 27 MHz band

(b) An interface circuit

(c) The decoder and

(d) Relay drive circuits

The receiver should be a crystal controlled superhet type, and should be sharply tuned to the transmitter frequency. Several such circuits in IC form are commercially available. Coupled to the receiver output is an interface circuit. This restores the original pulse shapes which get distorted in the receiver and also makes the pulses compatible with the TTL circuits that follow—both in amplitude and polarity.

This results in the output from the interface to be a replica of the original pulse train from the coder circuit in the transmitter, with the pulse widths maintained and the sync pulse in the correct position with respect to the other channels. The circuit of a suitable interface is given below.

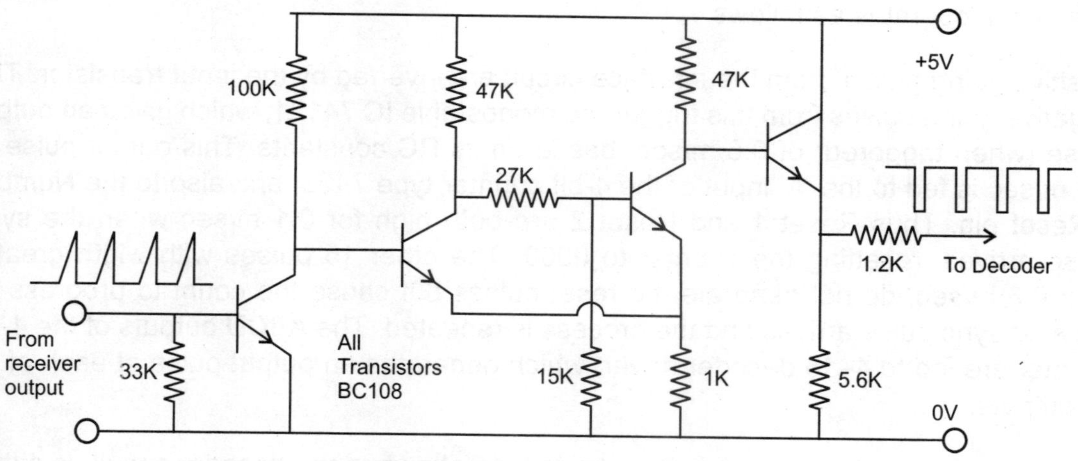

Figure 10.7 *Interface circuit*

The output pulses from the interface circuit are fed to the decoder. A circuit based on TTL devices is given below. This 74121 IC detects the sync pulse following which the 4-bit counter starts counting.

The ABCD outputs from the counter are connected to a 4-16 decoder driver, which provides (after the sync pulse), 15 output pulses to operate relays (in case of switched channels) or servo drives (for proportional channels).

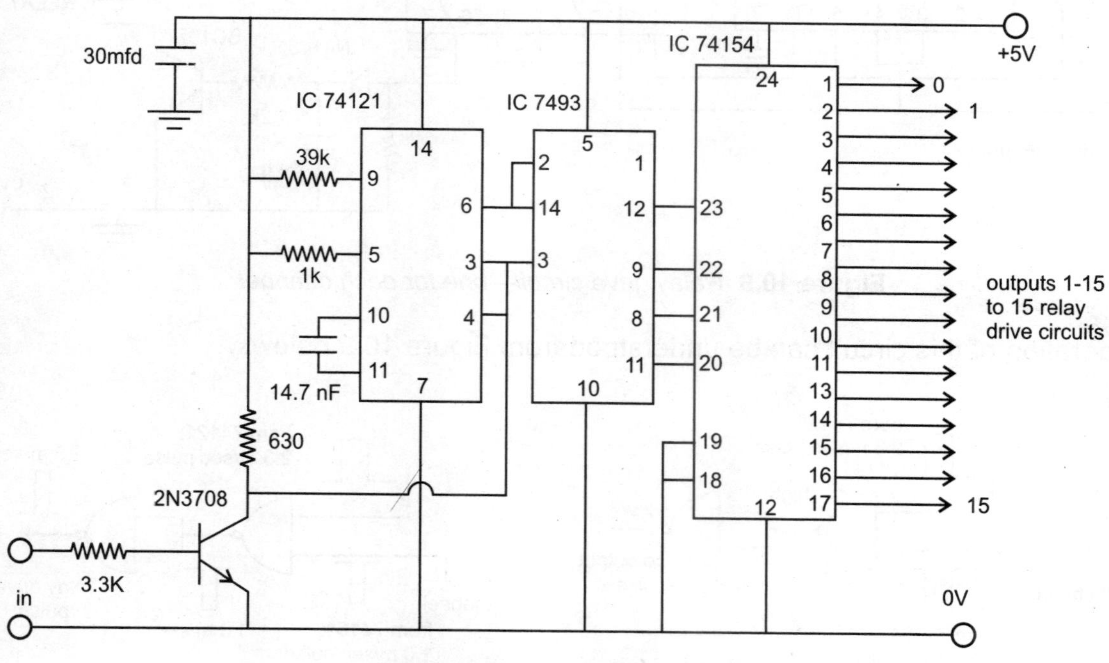

Figure 10.8 *Decoder-driver*

The circuit operates as follows:

Positive going pulses from the interface circuit are inverted by the input transistor. The negative going pulses from this trigger the monostable IC 74121, which gives an output pulse (when triggered) of 0.6 m/sec, based on its RC constants. This output pulse of 0.6 m/sec is fed to the 'A' input of the 4-bit counter type 7493, and also to the Number 2 Reset pin. Thus Reset 1 and Reset 2 are both high for 0.1 m/sec when the sync pulse arrives, resetting the counter to 0000. The other 15 pulses with width greater than 0.6 m/sec, do not generate the reset pulses but cause the count to progress till the next sync pulse arrives and the process is repeated. The ABCD outputs of the 4-bit counter are fed to 4–16 decoder driver, which generates an output pulse at each of its 16 outputs.

The width of the output pulses from outputs 1-15 shown in the decoder circuit, is either 2.3 m/sec (if the channel is inoperative), or 1 m/sec (if the channel is switched on). These pulses are fed to relay drive circuits as shown in Figure 10.8 below.

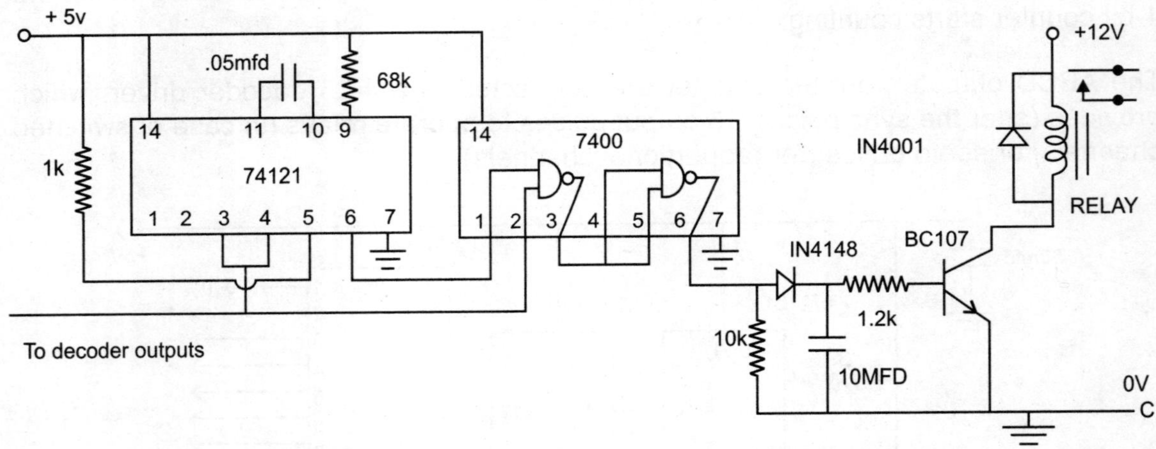

Figure 10.9 *Relay drive circuit—one for each channel*

Operation of this circuit can be understood from Figure 10.9 below.

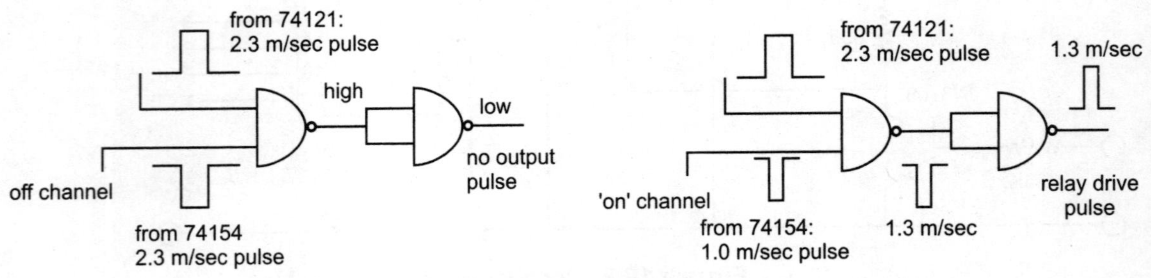

Figure 10.10 *Relay drive circuit's timing diagram*

The 74121 monostable is triggered by negative going pulses from the decoder IC 74154. With the RC constant shown, the 121 generates 2.3 m/sec pulse at its Q output (pin 6), which is connected to one input of the NAND gate 7400; the other input of the 7400 being connected directly to the 74154 output.

Hence there is no output from the gate when the pulse is more than that of a switched channel, which is 1 m/sec. The output from the 'on' channel is fed to the relay driver transistor BC107, which is diode protected. A capacitor of 10 Mfd connected at the input of the relay drive transistor prevents relay chatter, since the pulse width is for duration of only 1 m/sec in the pulse train.

It should be noted that for each of the 15 decoder outputs, a separate relay driver is needed (if the channel operation is required). The relay contacts should be wired for motor control or other requirements, keeping in view the contact ratings of the loads.

Fabrication of Printed Circuit Boards

Printed circuit boards or PCBs (as they are known) are the backbone of miniature electronic circuits. Replacing point-to-point wiring, printed circuits are the key to mass production of electronic devices.

To fabricate printed circuits, we need to start with the basic material—copper clad laminates. These consist of an insulating board on which copper sheet is pasted in the manufacturing stage.

The insulating boards in common uses are Hylam and fibreglass. Of these, Hylam is most common whereas fibreglass-based laminates are costlier but offer better thermal properties.

The copper cladding can be one sided or on both sides of the laminate. Double sided PCBs have the advantage that, with components mounted on both sides, the physical dimensions of the finished product is substantially lower than when single sided PCBs are used.

Once the copper clad board is selected, the second stage of the fabrication is to remove the unwanted copper and leave the circuit wiring on board. There are two common ways this could be done:

(a) Chemical method: Draw the required wiring on the board using an acid resistance ink.

(b) Mechanical method: Remove the unwanted copper physically using an appropriate tool.

Chemical Method

The circuit has to be drawn on the board using an acid resisting ink, which is available from electronics speciality shops. Duco or synthetic enamel paints are also acid resistant and could be used in place of acid resisting ink. When paints are used, a very fine paint brush has to be employed; otherwise the lines become very thick. Duco paint has the advantage that it can be sprayed on to the board, using a stencil carrying the pattern of the circuit to be produced. This method can also be used for small quantity production of PCBs. Once the etching is over, the paint can be removed using an appropriate thinner, leaving the copper lines exposed.

Once the lines are painted and completely dry, then the unwanted copper is etched away using a chemical which dissolves away the exposed copper on the board. Ferric Chloride solution is suitable for this process. The solution has to be heated.

Mechanical Method

This method requires special equipment and is computer controlled. The copper lines that are to be retained are programmed through a programmer and loaded into the instrument. A special tool is used to physically remove the unwanted copper from the laminate.

Though this method is superior to the chemical one, the equipment is costly and its use justifies only when large scale production is required.

12 Eddy-current Testing—A Non-destructive Method

Two methods are available for non-destructive testing of metal shells, namely ultrasonic testing and eddy-current testing. Of these, the eddy current method is more economical, reliable and capable of higher inspection speeds with simple equipment as compared to the ultrasonic method.

When a varying magnetic field is applied to a substance, currents are generated in it, which are called eddy-currents. This causes reduction of power which is converted to heat. In transformers and dynamos, eddy currents are prevented by using laminated cores instead of a solid block, each lamination electrically insulted from others by use of suitable lacquers.

One application of eddy currents is the induction heater, where heat is generated to cook food and another to melt metals in specially designed furnaces. In this chapter, we describe how eddy current technique can be used for non-destructive testing of metallic objects.

In the eddy-current testing of metal shells, the test object is placed in a varying magnetic field of a coil carrying an alternating current. Eddy-currents are induced in the test object by this alternating magnetic field. The inductive reactance of the coil is determined by the eddy-currents which create an additional alternating current magnetic field in the coil vicinity. The eddy-currents depend on the nature and condition of the test object. The coil impedance will change with the condition of the test object, if other parameters are held constant.

The factors which affect the coil impedance are:

(a) Electrical conductivity of the test object

(b) Physical dimensions

(c) Magnetic permeability and

(d) Presence of discontinuities, such as cracks or cavities

In any particular test, the factors (a), (b) and (c) above are held constant within close limits. Consequently, the presence of discontinuities contributes to change in the test-coil impedance, a measure of which indicates the condition of the sample under test.

Figure 12.1 shows the arrangement of the standard coil and the test coil in a bridge circuit. The two other arms of the bridge are made up of R & C components. The output of the bridge is fed to the electronic processing circuit, which consists of an amplifier, a rectifier and a trigger circuit—this is to operate an indicator.

The test frequency 'Fg' is a fundamental parameter in eddy-current testing. The limit frequency for tubes can be calculated for thin-walled tubes by the relation:

$$Fg = \frac{5066}{\mu rel\ \sigma\ di\ w}$$

Where:
- μrel = relative permeability of the test object
- σ = electrical conductivity of the test object in meters/ohm-mm
- di = inside diameter of the tube in cms
- w = wall thickness in cms

For aluminium shell samples tested:

$\mu rel = 1$, s = 35, di = 0.57 and w = 0.05

Substituting in equation above:

$$Fg = \frac{5066}{35 \times 0.57 \times 0.05} = 5078 \text{ Hz}$$

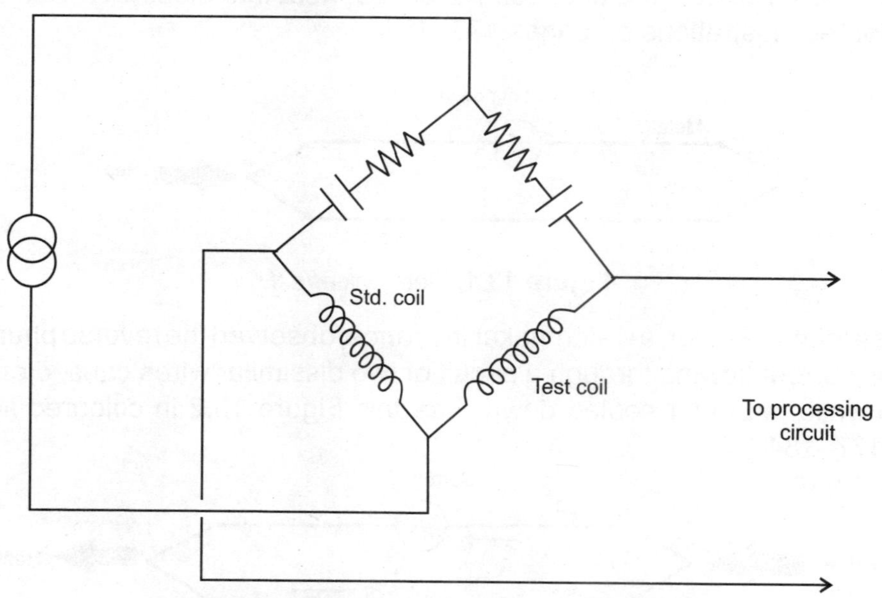

Figure 12.1 *Bridge circuit for eddy-current testing*

Frequency selection for crack detection

Thin walled tube are tested at frequency ratios of F/Fg = 0.4–2.4, the maximum sensitivity being at F/Fg = 1

The limits of 0.4–2.4 mentioned above are for a decrease of maximum test sensitivity to 71 per cent. In the present case, this sets the limit of test frequencies from approximately 2–12 KHz, with the maximum sensitivity at about 5 KHz.

13 Thermoelectric Phenomena

Thermoelectricity implies the direct conversion of heat energy into electrical energy and the transformation of electrical energy into heat.

The scientist who first observed the effect was Thomas Johann Seebeck (in 1821) and the phenomenon is called Seebeck Effect.

Seebeck observed that when a magnetic needle was held close to a pair of two conductors made of different materials (which were joined together at the ends), the needle was deflected when one of the junctions was heated.

In short, Seebeck Effect is the direct conversion of heat into electricity. See this Figure 13.1 in coloured illustrations on pages 177–184.

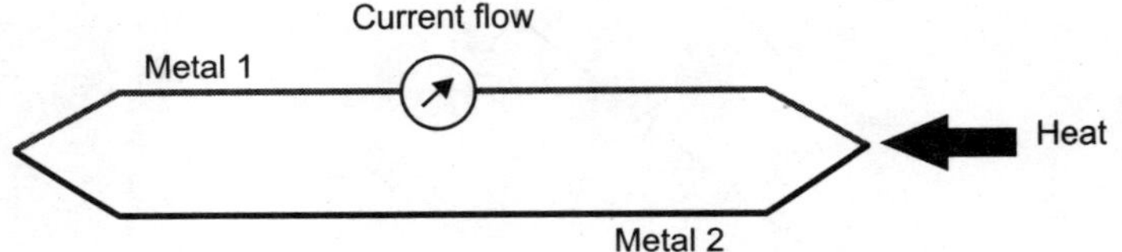

Figure 13.1: *Seebeck effect*

Thirteen years later, Peltier, a watchmaker in France, observed the reverse phenomenon–that electric current flowing through a circuit of two dissimilar wires caused one junction to heat up while the other cooled down. See this Figure 13.2 in coloured illustrations on pages 177–184.

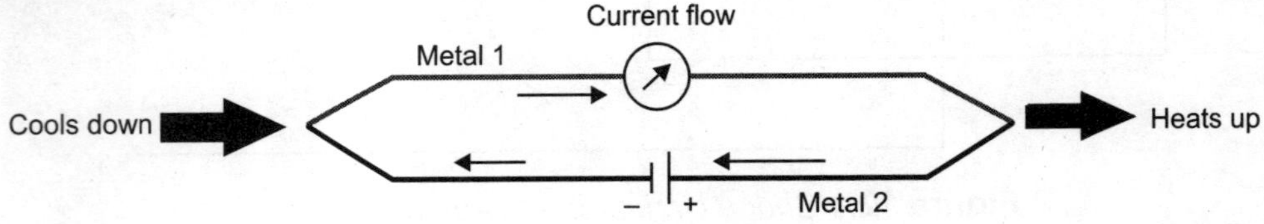

Figure 13.2: *Peltier effect*

Four years after the discovery of the Peltier Effect, Russian scientist E. Lenz, demonstrated how the latest discovery could be put to use. He put a drop of water on a thermo-electric junction, switched on the current which resulted in the water turning to ice. When the battery polarities were changed, the ice became water again.

Localised heating and cooling can be done using thermo-electric modules. This has important applications in electronics instrumentation. In a thermoelectric circuit, the voltage produced is due to two different phenomena—one of these is the Peltier Effect, described earlier and the other one is the Thomson Effect. When an electric current J passes between two points of a homogeneous wire whose temperature difference is ΔT, an amount of heat $\sigma J \Delta T$ is emitted or absorbed in addition to the Joule heat. Here, σ is the Thompson coefficient.

While the Peltier effect is due to the voltage which results from the contact of two different metals, Thomson effect is due to the temperature difference between any two points in a single conductor. Much later, in 1926, it was discovered that an oxidised copper plate conducts electric current in one direction but not in the other. This led to the conversion of AC into DC, a process commonly known as rectification. A parallel discovery showed that when light falls on such plates, an electric current is produced. These discoveries led to the development of semiconductors, such as transistors, solar cells, etc. Subsequently, it was found that thermoelectric effects in semiconductors are an order of magnitude higher than that in metals.

When one end of a semiconductor is hotter than the other, migration of electrons towards the cold end is more than the reverse flow. The cold end becomes negatively charged with respect to the hot end. When a p-type semiconductor and an n-type semiconductor form a p-n junction, Seebeck Effect is observed. See this Figure 13.3 in coloured illustrations on pages 177–184.

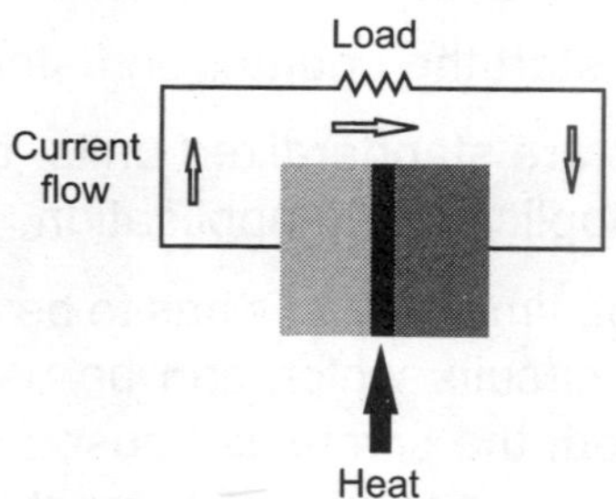

Figure 13.3 *Effect in a p-n junction*

Such semiconductors can transform energy the other way also. If electrical energy is supplied to the semiconductor instead of heat, current flowing through the device will heat one end and cool the other. Thermoelectric coolers and heaters operate on this principle. This technology has already yielded many products like refrigerators and heaters as well as cooling modules in electronic equipment. In such applications, the heat can come from any source, such as kerosene lamps and even sunlight.

One of the major applications of this technology is that changing the direction of current flow, just by operating a single switch, causes the refrigerator to also act as a heater.

14 Measurement of Very Short Time Intervals

Chronographs are instruments that record time intervals very accurately. A stop watch is an example of a chronograph. For very short time intervals, a stop watch is of no use since its operation depends on the response time of humans, which is more than a second of time.

Alternate methods of short time intervals have been available since many decades, with their accuracy much improved since the advent of semiconductor devices. We will discuss here a technique for measuring minute time intervals (in microseconds), using what is known as the counter-chronograph technique.

A counter chronograph has three essential components:

(a) A high frequency oscillator which generates pulses of the desired frequency

(b) A digital counter (with a readout) to record and display the number of pulses received during the time interval in which the chronograph is active

(c) A start-stop mechanism to start the counter and stop it during the measurement

While the first two components are standardized units, the third one, which is the start stop mechanism, varies from application to application.

The oscillator which provides the timing pulses has to be crystal-controlled for accuracy. There are several varieties of circuits which can be used for this application. Where very high accuracies are desired, the crystal is housed in an oven whose temperature is kept constant. For most applications, this refinement is not necessary.

Oscillators are driven by transistors, ICs or microcontroller devices as per the specific requirements.

Digital counters with capability of counting pulses at several MHz frequencies are now available. Thefcy are mostly based on ICs. VLSI devices (Very Large Scale ICs) combine the operations of counting and display in one package. For the projects under consideration in this book, discrete devices are utilised due to their easy availability and low cost.

The count output has to be presented through a display. This can take the form of LCD devices, LED packages or neon based displays. Whatever the type used, the output has to be in digits.

The crucial part of the instrument is the start-stop mechanism. This will depend on the application at hand but all have a common feature. The start stop pulses operate the gating circuits in the chronograph and are generally based on flip-flop circuits. These, in turn, are switched through mechanical, ultrasonic, optical or some other kind of signals and are application specific.

When a start signal is received from the start flip-flop, the counter start counting the pulses from the oscillator till the stop signal is received. The duration in between the start and stop signals, is presented on the readout used with the counter.

If the start-stop signals are to be activated by sound, then crystal microphones are the triggering mechanism. If optical signals are to initiate the timing process, then opto-electronic devices act as the input triggers.

The accuracy of measurement depends on the speed at which the start/stop signals will initiate the counting process. Hence, the input circuits which interface with the transducers need to be fast acting.

A block diagram of the counter chronograph is given in Figure 14.1.

Applications of counter-chronograph include the measurement of sonic velocities in small samples of solids, the measurement of bullet velocities (described under Major Projects in this book), even measurement of the speed of light. In civil engineering, it has applications in the non-destructive testing of concrete. Measurement of camera shutter speeds has been carried out using the counter-chronograph instrument. The flexibility of this instrument lies in its ability to perform a wide variety of tests, with accuracies as high as desired.

Decay times of phosphors (in the millisecond time region) have also been carried out using a counter-chronograph with photo-electric transducers to start and stop the counting process.

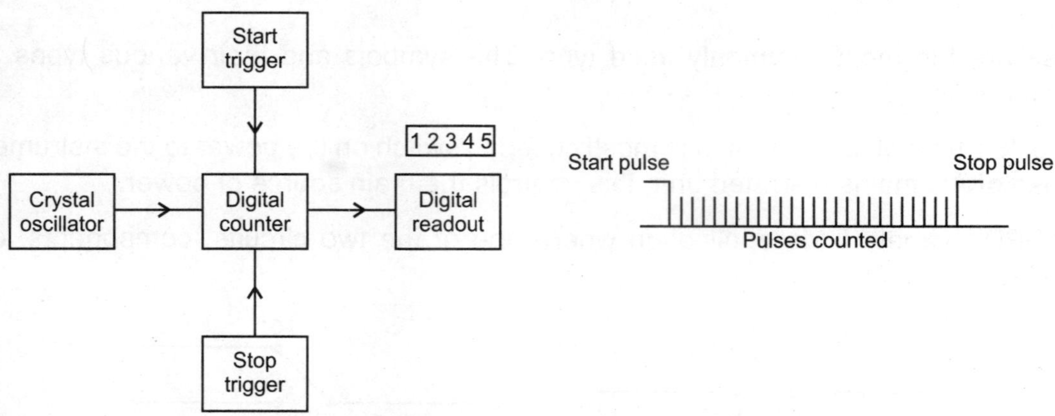

Figure 14.1 *Block diagram of counter-chronograph*

PART II

Components and Special Chemicals

15 Electrical Switches and Electromagnetic Relays

All instruments which operate on electric power utilise some form of switches. They may be used to switch power on/off, select a particular component or determine which of the available voltages are required.

There are basically four types of switches normally used in test and measuring instruments. These are:

(a) Toggle switches

(b) Rotary switches

(c) Limit switches

(d) Slide switches

We will confine our discussion to the first three types used in our projects.

Toggle Switches

These are the most commonly used type. The symbols and their various types are shown below.

The SPST type of toggle switch is mostly used to switch on the power to the instrument. In case of AC mains operated unit, this controls the main source of power.

The SPDT variety finds application where one of the two circuits, components, etc. needs to be selected.

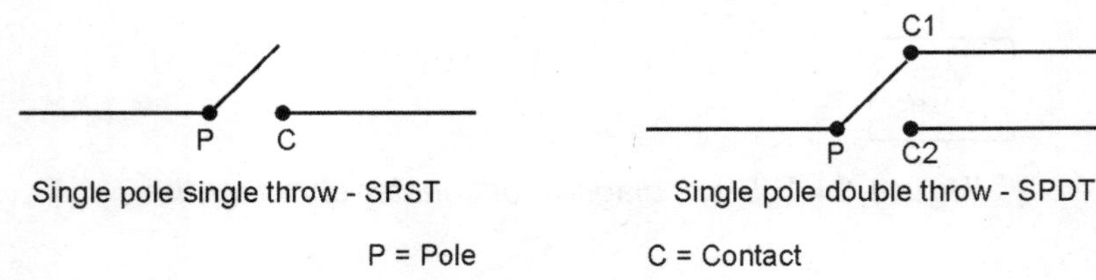

Figure 15.1 *Single pole toggle switches*

Where two input lines are to be simultaneously switched on/off, the DPST type is required. This is shown in Figure15.2 below.

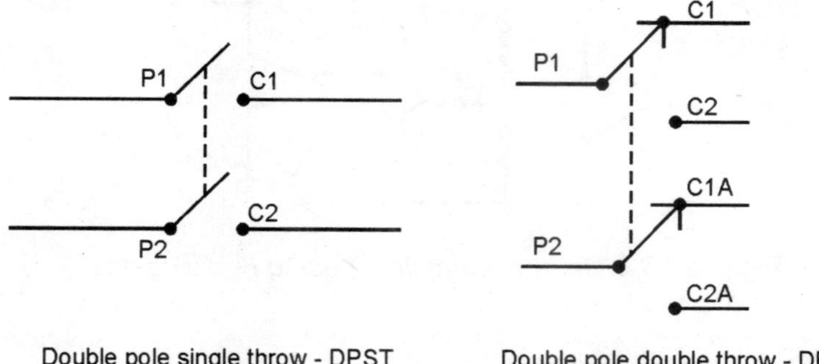

Figure 15.2: *Double pole type of toggle switches*

The DPDT type of toggle switch finds application in circuits where two lines have to be connected to one of the two different contacts, from where other components are wired. Most of the toggle switches can be used with voltages commonly encountered in electronic circuits. However, current ratings of the contacts are important. When switching higher currents than what they are rated for, permanent damage can result to the switches.

Mechanical dimensions of the switches have to be relevant to the environment in which they are used. Many problems arising in IC circuits are due to electromechanical devices; switches being no exception. Most switches have the contact bounce effect shown below:

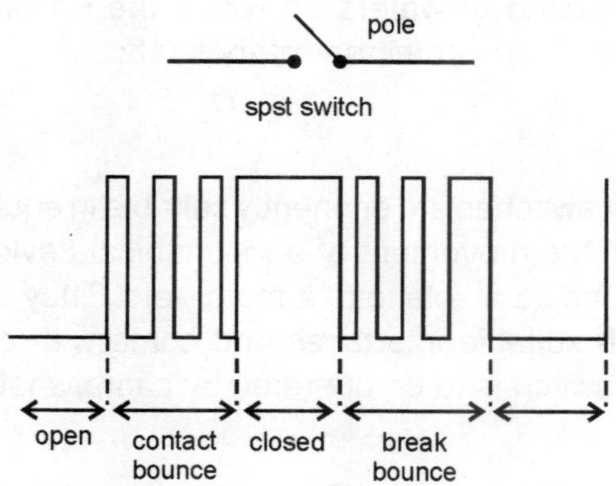

Figure 15.3 *Contact bounce effect in mechanical switches*

If these switches are used directly in logic circuits, de-bouncing circuits should be used. One such circuit using NAND gates is shown below. Costlier micro-switches (of which several varieties are available) are preferable. However, it is not always possible to replace toggle switches with micro-switches due to the difference in physical structure and mode of operation between the toggle switches and micro-switches.

Micro-switches or limit switches are described later in this chapter.

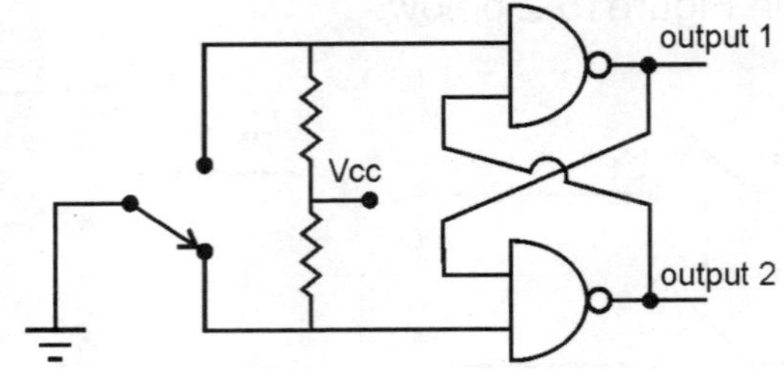

Figure 15.4 *De bouncing circuit using NAND gates*

Rotary Switches

Rotary switches come in many varieties, from single pole two-way to multi-pole, several way types. Their sizes vary from miniature to very large, capable of handling a vast range of voltages and currents.

Multi-pole multi-way rotary switches have applications where other types cannot be used.

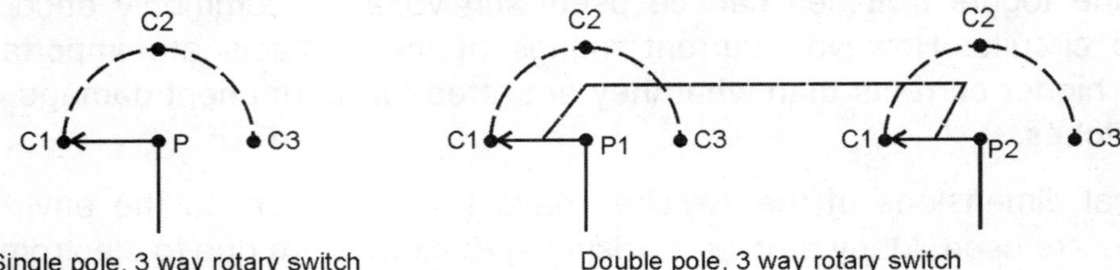

Figure 15.5 *Single pole and double pole rotary switches*

These switches are composed of wafers on which the poles/contacts are fixed. The wafers themselves are fixed on a common rotary shaft.

Limit Switches

As the name signifies, limit switches are eminently suitable in equipment where a switching action is to be linked with the movement of a mechanical device (either to switch on or switch off an actuator), such as a solenoid, a motor, etc. They are also utilised to switch electronic circuits through suitable interfaces and come with or without rollers. Rollers come in handy when the switch is to be operated by cams, shafts or disks.

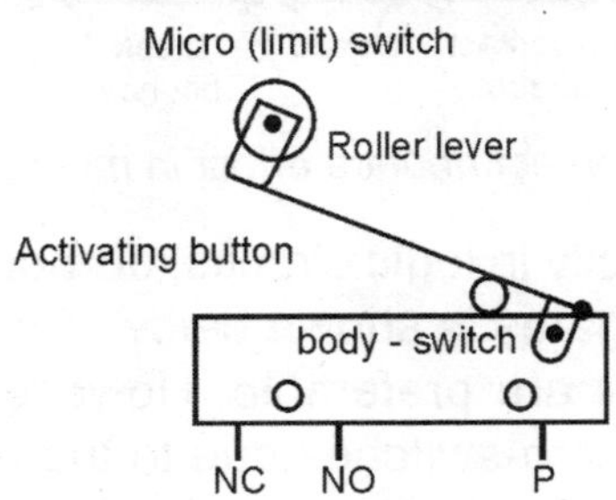

Figure 15.6 *A micro (limit) switch with roller*

Electromagnetic Relays

Relays are most usefully employed when a number of independent circuits are to be switched on from one signal. In this aspect, multi-pole relays have an important advantage over power transistors. A further advantage is the switching of AC circuits. However, in some circuits used, we find that the relay itself is switched on through a transistor. Thus, the transistor base current is all that is needed to be supplied from digital logic—an important aspect in consideration of loading factors.

Relays used with transistors in this manner should have protective diodes across them as shown below.

When current through the relay is switched off, its rapid decay through the relay coil generates a back emf, which is of several times the magnitude of the supply voltage, which can damage the drive transistor.

The clamping diode offers a low impedance path to this spike and protects the transistor.

A problem encountered with relays switching heavy loads is contact arcing. The arc generates 'glitches' which travel through the bus line and upsets the logic levels of ICs. This can be solved by the use of an opto-isolator between the logic circuits and relay. The contact rating of relays should not be exceeded. Hermetically sealed relays and reed relays are also available and should be used wherever their use is relevant. Hermetically sealed relays are enclosed in an airtight compartment and do not permit moisture to affect the relay contacts, thus avoiding corrosion of the metallic parts. Reed relays are a special brand of devices, which are extensively used in communication systems.

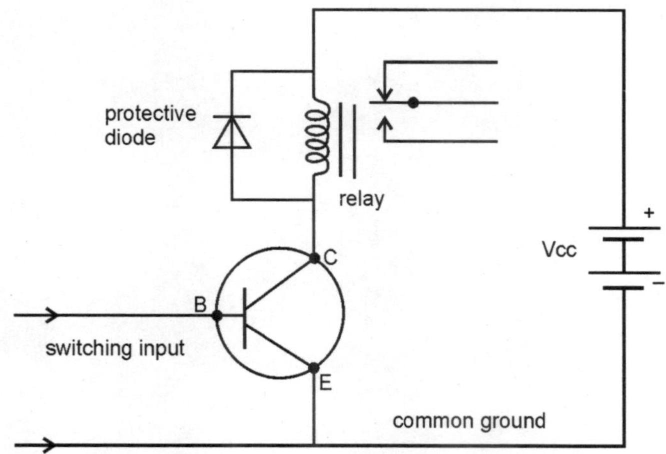

Figure 15.7 *Transistor controlled relay with protective diode*

Reed relays consist of two flat reeds with special tips, which are sealed in a glass envelope. The glass enclosure is either evacuated to remove air or is filled with an inert gas like nitrogen. In either case, there is no oxygen inside the glass enclosure; hence corrosion of the contacts is avoided.

The reeds are made of ferromagnetic materials and the tips are mercury coated to provide low contact resistance. In the absence of a magnetic field, the reed contacts are open. When a magnetic field is applied, the contacts close and the relay circuit is completed.

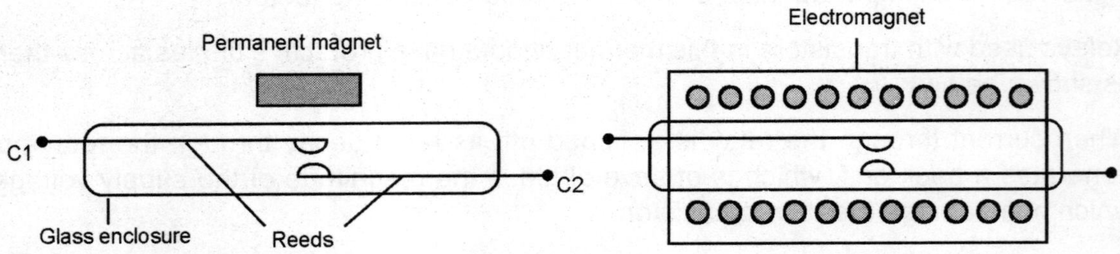

Figure 15.8 *Reed relays with permanent and electromagnets*

The magnetic field can be applied either by a permanent magnet or an electromagnet. In the former case, the permanent magnet has to be brought near the reed relay by a mechanical action, linking the switching action to a mechanical movemcnt (compare this with the limit switch which is also linked to a mechanical movement).When switching action is required through electrical pulses, the reed relay is actuated by an electromagnet. Both types of relays are illustrated in Figure 15.8 above.

In communication systems, fast acting reed relays are employed quite commonly. The inert atmosphere and the mercury wetted contacts ensure that contact bounce (common with toggle switches) is not encountered.

Passive Electronic Components

16

In this chapter, we will consider passive electronic components, which cover resistors, capacitors and inductors.

Resistors

Resistors can be categorised according to their composition and whether they are of fixed ohmic value or variable type. Their specifications will cover their ohmic value, the wattage and tolerance (in terms of percentage accuracy).

Composition of Resistors

The three basic types are carbon based resistors, metal film type and wire-wound resistors. In the carbon based, there are also sub types—the carbon composition and the carbon film type.

While the carbon and metal film resistors are generally of low power rating (up to a few watts), wire wound resistors are available in several watts rating. The power ratings also apply to variable resistors, such as, carbon potentiometers having a low power rating while the wire-wound types going up to 100 watts and more.

Wire-wound resistors which are circular also have an inductive component when used in high frequency circuits. The most common type of carbon potentiometers are single turn, while wire-wound potentiometers of ten-turn variety are in common uses.

In specific applications, fixed resistors can be connected in series, increasing the effective resistance, while parallel connected resistors decrease the resistance.

The wattage dissipation of resistors is an important factor to be taken into account. Generally, for carbon based resistors, the heat dissipated (in terms of watts) should be less than 50 % of the rated wattage of the resistor.

Some useful formulae are given below:

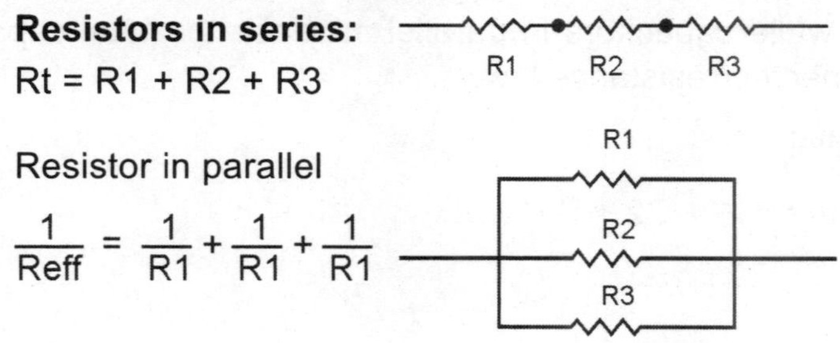

Resistors in series:

Rt = R1 + R2 + R3

Resistor in parallel

$$\frac{1}{Reff} = \frac{1}{R1} + \frac{1}{R1} + \frac{1}{R1}$$

Wattage dissipation = I^2R or $V \times I$

Where I is the current in amps through the resistor in ohms and V the voltage across the resistor.

Carbon resistors are colour coded for their ohmic value and percentage tolerance:

Colour	Ohmic value	Colour	Ohmic value	Colour	Ohmic value
Black	0	Orange	3	Blue	6
Brown	1	Yellow	4	Violet	7
Red	2	Green	5	Grey	8
White	9				

The colours are in the form of bands on the resistor. The first band is the first digit, the second one is the second digit while the third is the multiplier (number of zeros).

Tolerances are also colour coded, in form of a separate band, with percentages as follows:

Brown: 1%, **Red:** 2%, **Golden:** 5%, **Silver:** 10%, **None:** 20%

An example of colour-coded carbon resistor is given in Figure 16.1 in the coloured illustration.

Capacitors

Two metal plates separated by an insulating medium are what basically comprise a capacitor. The insulating medium termed dielectric – may be air, mica, polyester, electrolyte, ceramic, tantalum oxide, oil, etc. The capacity is generally specified in microfarads 10^{-6} or picofarads 10^{-9}, since a farad is a very large quantity and is not a practical unit.

Capacitors, like resistors, can have a fixed value or be variable. Variable capacitors have air as the dielectric while the fixed ones have the dielectrics mentioned above.

Apart from the rated capacity, capacitors are also specified for a maximum working voltage, to avoid puncturing the dielectric due to excess voltages. As in the case of resistors, they can be connected in series or parallel. However, the maths here is different. Capacitors in series yield a net lower capacitance following the same formula as for resistors in parallel, while capacitors in parallel increase the total capacitance according to the series connected resistance law.

Thus, for capacitors in parallel:

$$Ct = C1 + C2 + C3 \ldots\ldots$$

For capacitors connected in series:

$$1/Cnet = 1/C1 + 1/C2 + 1/C3 \ldots\ldots$$

These arrangements are shown in Figure 16.2 below:

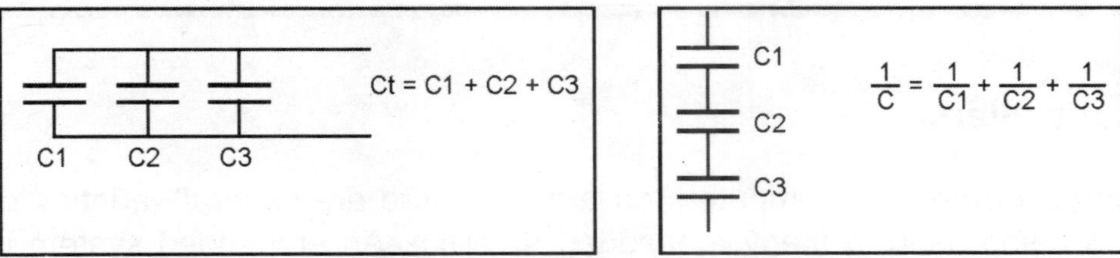

Figure 16.2 *Arrangement of capacitors*

Their application ranges from tuned circuits for frequency selection, time delay generation in RC circuits, blocking DC while allowing AC to pass through, etc.

Electrolytic capacitors offer a much larger capacitance as compared with other capacitors of the same physical dimensions. They are used mostly in power supplies to filter out the fluctuations in DC output due to the AC ripple following rectification. Tantalum capacitors are a type of electrolytic capacitors. They consist of a pellet of tantalum metal which forms the anode; this being covered by the insulating oxide layer, which is the dielectric and a conducting material which will be the cathode.

These capacitors have a higher capacity per volume and weight than the other types.

Tantalum capacitors have a lower equivalent series resistance, lower leakages, and can operate at higher temperatures than other electrolytic capacitors.

17

Active Devices

Microcontrollers

Microcontrollers are mini-computers on a chip. There are several varieties of micro-controllers being used in many embedded systems. An embedded system is where a microcontroller is used to perform only one task. In the domestic environment, embedded systems are TVs, VCRs, Camcorders, etc. In communication systems, embedded devices include cell phones, intercoms, etc. In offices, we have examples like fax machines and printers. Domestic appliances like sewing machines have a microcontroller which is programmed to do just what the machine is meant for—sewing clothes. The automobile industry has several examples of embedded products including engine control ones. Embedded systems also rely on various types of sensor to a large extent.

In this section, we will outline a popular type of microcontroller—the 8051 which is a 40 pin, 8-bit device.

This micro has 2 timers, 4 I/O ports accessible through 32 pins, 1 serial port, 4k bytes of ROM, 128 bytes of RAM, 3 timers and 8 interrupt sources. The speeds of the device vary from 12–20 MHz, depending on the exact part number.

In contrast with microcontrollers, microprocessors such as 8085 (Intel), 6800 (Motorola) and 6502 (Rockwell) do not contain on-board I/O ports, RAM, ROM and Timers. These have to be provided for separately to form a complete system for specific applications.

The pin-out diagram of this microcontroller is given in Figure 17.1.

Pin Functions

In the 8051, pins 1–8 are assigned for port 0. Pin 9 is the Reset pin. This is normally low and is active high. When a +ve pulse is applied here, the 8051 terminates all activities and is reset. The duration of the pulses should be at least two machine cycles for the micro to be reset.

Pin name	Pin	8051	Pin	Pin name
P1.0	1		40	Vcc
P1.1	2		39	P0.0
P1.2	3		38	P0.1
P1.3	4		37	P0.2
P1.4	5		36	P0.3
P1.5	6		35	P0.4
P1.6	7		34	P0.5
P1.7	8		33	P0.6
RST	9		32	P0.7
P3.0	10		31	EA
P3.1	11	8051	30	ALE
P3.2	12		29	PSEN
P3.3	13		28	P2.7
P3.4	14		27	P2.6
P3.5	15		26	P2.5
P3.6	16		25	P2.4
P3.7	17		24	P2.3
XTAL2	18		23	P2.2
XTAL1	19		22	P2.1
GND	20		21	P2.0

Figure 17.1 *Microcontroller 8051 pin-out*

What is a Machine Cycle?

In the 8051, one machine cycle requires 12 oscillator periods. Thus, for 'reset' operation we need at least 24 oscillator periods. If the clock frequency is 20 MHz, each oscillator period is 0.05 micro-second; hence for two machine cycles, the duration of the reset pulse should be at least 24 × 0.05 microsecond or 1.2 microseconds. For lower clock frequencies, the duration of the reset pulse will be correspondingly higher.

Pins 10–17 are port 3 pins. Like other 8051 ports, this is also an I/O port. The 8051 needs an external clock to run the oscillator in the chip. A crystal oscillator is connected between pins 18 and 19, marked XTAL 2 and XTAL 1 respectively. An external oscillator can also be used, in which case, it is connected to pins 19 (XTAL 1), pin 18 being left unconnected. Pin 20 is the ground pin. The negative side of the power supply (+5 V) is also connected here. Pins 21–28 connect to port 2.

Pin 29 is **$\overline{\text{PSEN}}$**- Programme Store Enable. When an external ROM is used to hold the programme code (in 8031 systems), this pin is used. Pin 30 is ALE—Address Latch Enable, for connecting to external memory as in 8031 systems. Pin 31 is the External Access pin. In the 8051, which has an internal ROM, this is connected to Vcc. In 8031 for ROM located outside the device, this is to be connected to ground.

Pins 32–39 are to access port 0, which has open collector outputs. These outputs must be connected to Vcc through pull-up resistors of 10k each.

Finally, pin 40 is for the Vcc, +5V supply.

Registers and Stack

The 128 bytes of RAM in the 8051 are utilised as follows: Ram locations 00-1F (Hex) are for register banks 0-3. In this Register Bank 1 occupying RAM location from 08-0F is also used for stack. Ram locations 20-2F are bit addressable while those from 30-7F are for scratch pad RAM.

Assembly Language

The CPU in the microcontrollers (as well as in microprocessors) can only function in the binary system; i.e. a set of 0s and 1s. This is called the machine language. However, to run programmes in machine language is a tedious affair; thus programming is carried out in higher level languages like, 'C', etc. When a programme is written in higher level languages, it has to be converted into machine language for the microcontroller to understand this and function. This is implemented by using a compiler which converts the higher level language into the machine language.

Some applications of microcomputers in embedded systems are given in Projects in Part 4 of the book.

Opto-Electronic Devices

Light-emitting Diodes

The most widely used opto-electronic device is the light-emitting diode, or LED. LEDs are semiconductor devices and, when suitably forward biased, emit light of different colours, depending on the specific LED in use. Common colours are red, green, yellow, amber and blue. Combinations of different colours in a single package yield a light of white colour. Infra red LEDs are also available. These find extensive applications in remote controls for T V and other devices.

Compared to other sources of light, LEDs have some distinct advantages. They operate cool and can be used for more 10,000 hours.

Their current consumption is low and have fast switching response – ideal for digital circuit applications. Although the earlier LEDs had a low light output, new developments have offset this disadvantage to a considerable extent. For instrumentation and interior displays, they are unparalleled.

LEDs come in various sizes, from the smallest having a diameter of 1 mm to 10 mm and growing larger as new developments take place. While most LEDs have a circular cross-section, LEDs with rectangular cross sections are also available. When these devices need to be closely spaced, rectangular ones are to be preferred. Since they emit only one colour, combinations of these in a single package result in multi-coloured emissions. Such combinations have LEDs which are connected in parallel, with either the anodes or the cathodes connected together.

When LEDs of different colours are wired in parallel, they should have the same operating voltage, otherwise the LED with a higher operating voltage will not function due to the supply voltage being 'clamped' at a lower level. One way to avoid this is to use separate resistors in the anode circuit, as shown in the circuit below.

The active semiconductor wafers which produce the light are encapsulated in a plastic material which is very sensitive to high temperatures. If the lead wires from the encapsulation are overheated while soldering, the plastic covering melts and the LED is rendered useless.

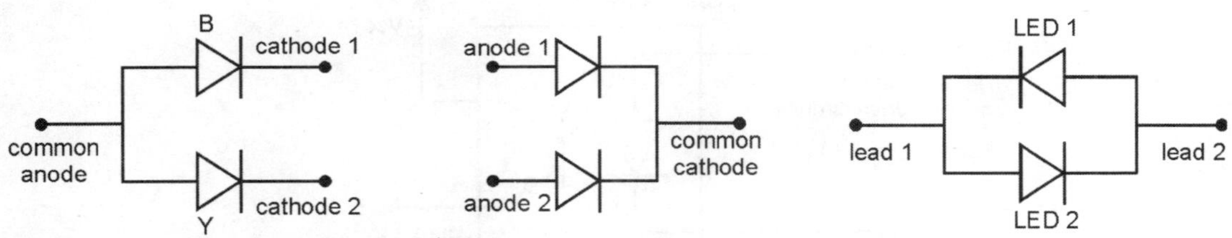

Figure 17.2 *Electrical connections in multiple LED packages*

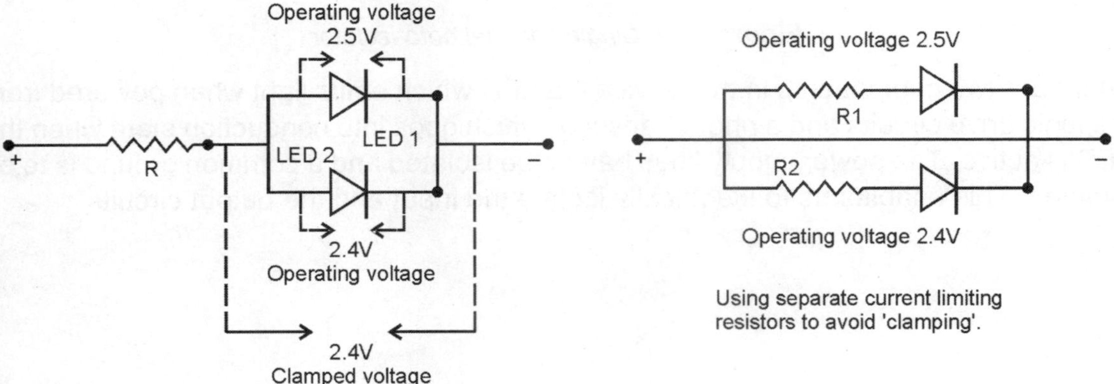

Figure 17.3 *Paralleled LEDS: Clamping and a solution*

The low operating voltage of LEDs renders them safe to use in hazardous environments. When displays comprising several LEDs are fabricated, multiplexing techniques are useful since they reduce the number of ICs in the overall circuit.

Opto-isolators

We now introduce an opto-electronic device which is known as an Opto-isolator. When logic circuits have to interface with other high power devices like power MOSFETs, at times problems may arise as they cannot be connected directly. DC motors are driven either through power semiconductors or through relays which are driven by relay drive circuits. Electromagnets and DC motors generate electrical disturbances which travel through the voltage supply lines and reach the logic circuits. Known as glitches, these noise signals are of very short duration. They are generated during the 'switch off' instant of the electromagnetic devices, since the current changes peak at such times. A device known as opto-isolator will solve such problems. Opto-isolators have two active components in a single light-tight package. One of these is a LED which connects to the logic drive circuit, while the other is a phototransistor which receives the optical signal from the LED and connects to the power drive circuit. Since the two components are connected to different power supply lines, electrical isolation is achieved, the signal transfer taking place via the optical route.

An opto-isolator is shown below.

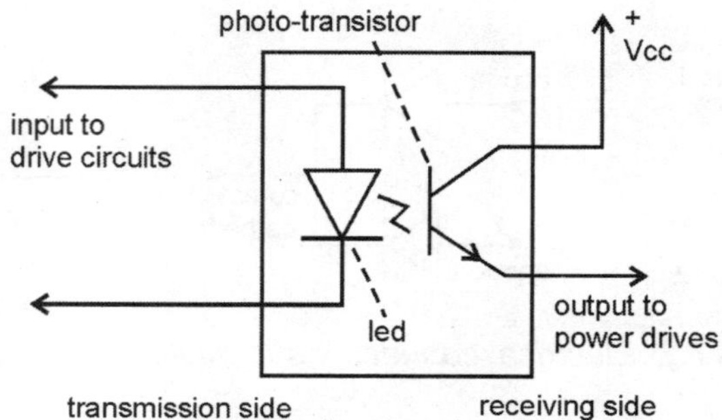

Figure 17.4 *Single channel opto-isolator*

There are two components in this device – a LED which emits light when powered from the logic drive circuits and a photo transistor which goes into conduction state when the LED is active. The power supply lines haveto be isolated and a common ground is to be avoided. This enables us to electrically isolate the input and the output circuits.

Electronic Sensors

18

Temperature Sensing Elements: Thermistors

One of the great advantages of using a thermistor sensor in an electronic thermometer is the rapidity with which temperature measurements can be made.

Thermistors: **Symbol:**

Types: Negative temperature coefficient: As temperature increases, its resistance falls:

TH –B12: Encapsulated bead type

VA 1098: R ~ 4.7 k ohms at 25 °C: Problem of insulating the leads.

Sensor: National type: LX5700 has built in op-amp and circuit:

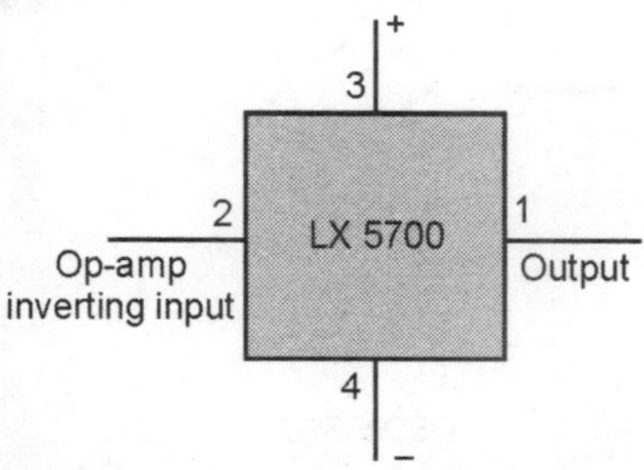

Figure 18.1 *Sensor connection diagram*

Analogue Devices: AD 590: 1 °C rise increases output current by 1 uA

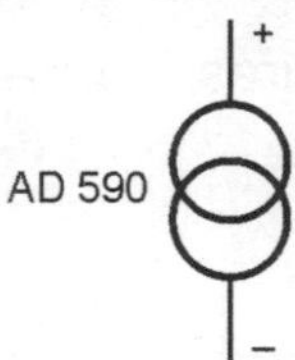

Figure 18.2 *AD 590 is a two terminal device*

Silicon diodes as temperature sensors:

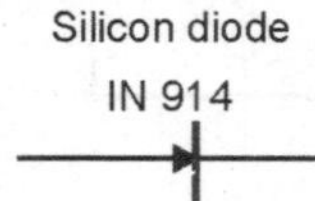

Figure 18.3 *Semiconductor diode*

Voltage drop across the diode junction changes @ 2.24 –mV/°C.

National Semiconductor Temperature Sensor, current output is 1 uA per degree Kelvin.

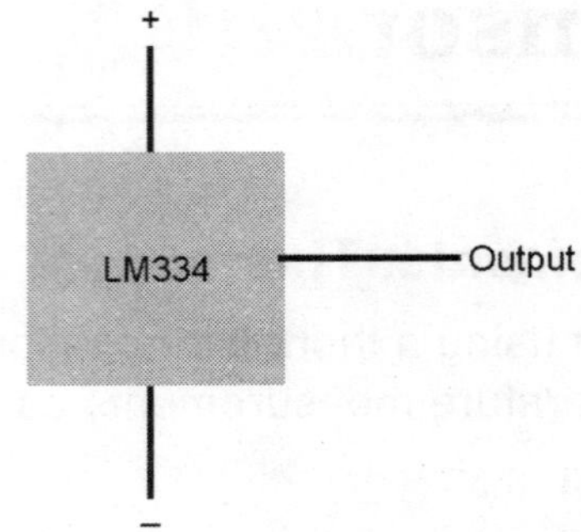

Figure 18.4 *Current varies with temperature in this sensor*

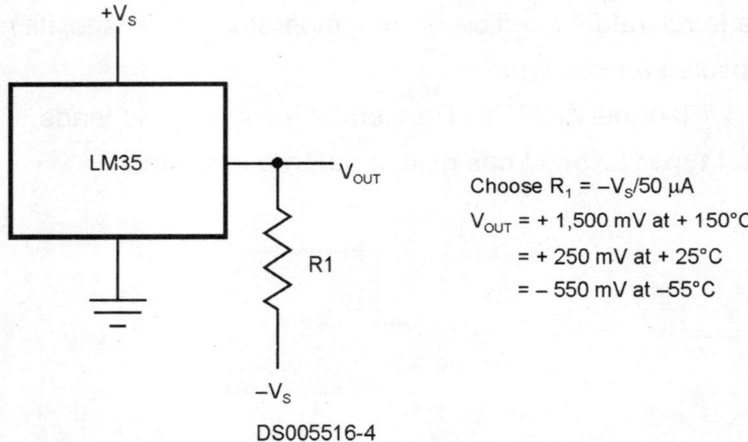

Figure 18.5 *Voltage varies with temperature in LM35*

Humidity Sensors

Figaro Electronics type NH-3: Requires 2.5 VAC at terminals 1, 3.

Resistance Rh changes with humidity

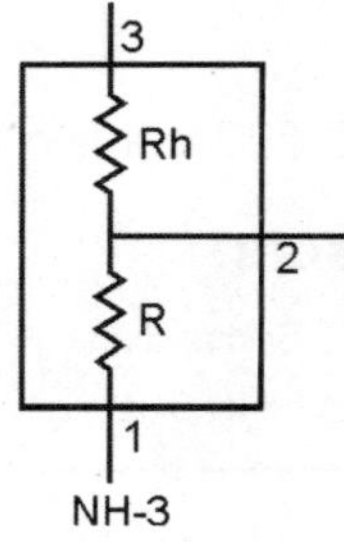

Figure 18.6 *Humidity sensor provides resistance change*

Phillips type H-1: Capacity changes with humidity.

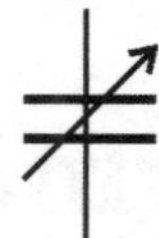

Figure 18.7 *Type H-1 sensor*

Gravity Force Sensors: Analog Devices ADXLO5

This sensor will detect a relative gravitational force of 0 g when positioned horizontally. It has a sensitivity of ~200 mV/g.

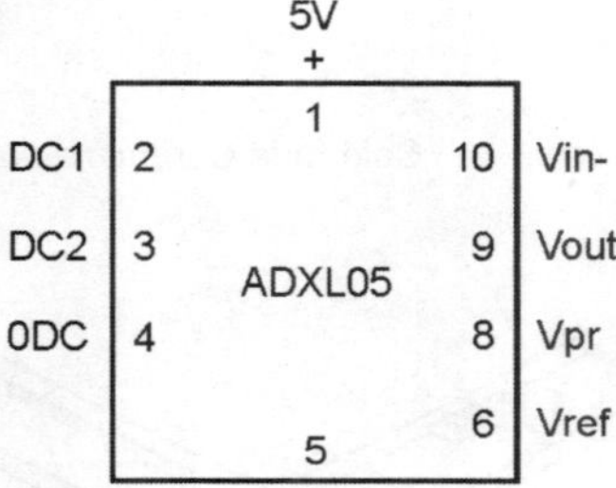

Figure 18.8 *Pin-out of Gravitational Force Sensor*

Hall Effect Sensor

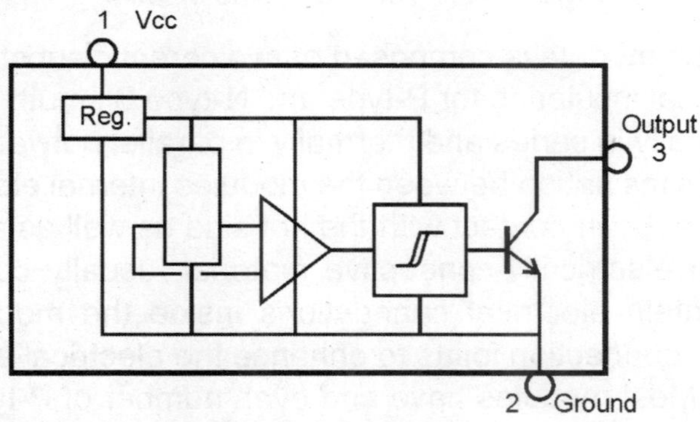

Figure 18.9 *Type OHN3040U from TRW incorporates a Hall element, a linear amplifier, a Schmitt trigger and a band-gap voltage regulator.*

Gas Sensor

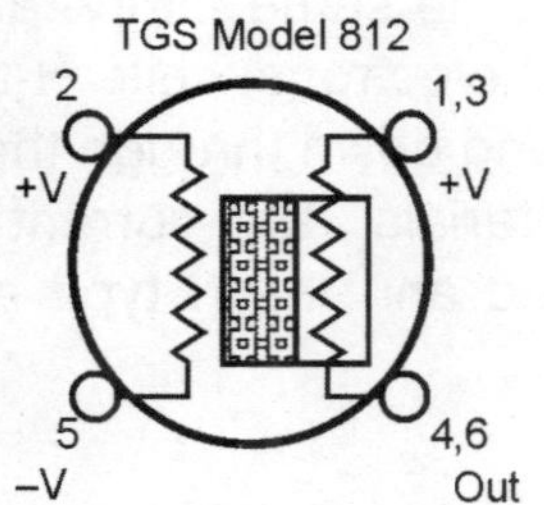

Figure 18.10 *Gas and Fume detector by Taguchi*

19 Solid State Thermoelectric Modules

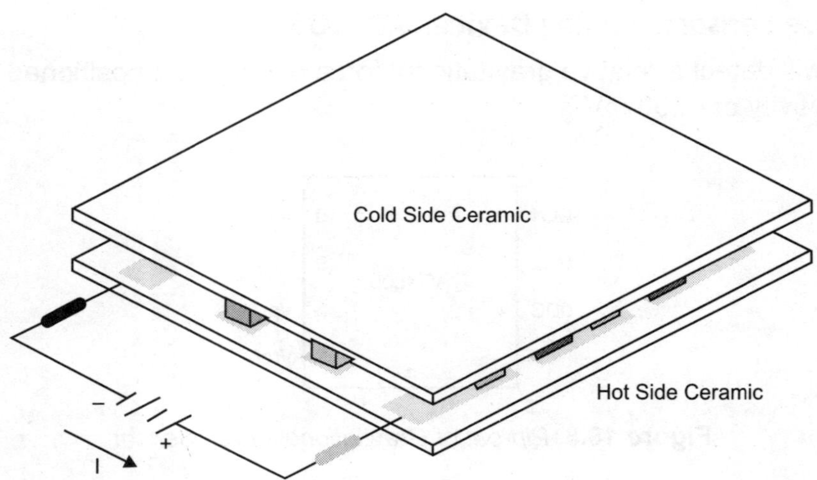

Figure 19.1 *Thermoelectric Module*

A typical thermoelectric module is composed of two ceramic substrates that serve as a foundation and electrical insulation for P-type and N-type Bismuth Telluride dice, which are connected electrically in series and thermally in parallel between the ceramics. The ceramics also serve as insulation between the modules internal electrical elements and a heat sink, which must be in contact with the hot side as well as an object against the cold side surface. An electrically-conductive material, usually copper pads attached to the ceramics maintain electrical connections inside the module. Solder is most commonly used at the connection joints to enhance the electrical connections and hold the module together. Most modules have and even number of P-type and N-type dice; one of each sharing an electrical interconnection is known as a couple. The above module would be described as an 11-couple module.

While both P-type and N-type materials are alloys of Bismuth and Tellurium, both have different free electron densities at the same temperature. P-type dices are composed of material having a deficiency of electrons while N-type has an excess of electrons. As current (Amperage) flows up and down through the module, it attempts to establish a new equilibrium within the materials. The current treats the P-type material as a hot junction, needing to be cooled and the N-type as a cold junction needing to be heated.

Since the material is actually at the same temperature, the result is that the hot side becomes hotter while the cold side becomes colder. The direction of the current will determine if a particular die will cool down or heat up. In short, reversing the polarity will switch the hot and cold sides.

Leads to the modules are attached to pads on the hot side ceramic. If the module is sealed, you can determine the hot side without applying power. With the module on a flat surface, point the leads toward you with the positive lead, usually in red wire insulation, on the right. The bottom surface will be the hot side.

Specifications of two thermoelectric modules are given in the below table.

ambient = 30C					
Item Part No	Description	Max Delta T	Max Voltage	Max Current	Max Wattage
TEC1-12705	Single stage TEC	68	15.4	5.5	49
TEC1-12708	Single stage TEC	68	15.4	8.5	76

The maximum voltage that can be applied to these two types is 15.5 V and the maximum current drawn at these voltages is 5.5 amps for the 12705 and 8.5 amps for the 12708.

20 Insulating Materials and Other Chemicals

Electrical insulators take the form of solids, liquids and gases. In liquids, insulating oils are available while in gases, dry air is also a good insulator. Solid electrical insulators in common use for several decades were glass, porcelain, mica, Bakelite, paraffin and rubber. Where mechanical and thermal properties were also of importance, porcelain was one of the best alternatives.

When insulators in the form of sheets were required, glass, mica, rubber and Bakelite were most suitable, due to the ease with which the sheets could be fabricated.

The main parameter to be considered in insulators is the maximum voltage that the material can withstand without breaking down, specified in V per unit thickness, of the insulating material.

During the past fifty years, several synthetic materials have been developed which act as good electrical insulators. These are polystyrene, Perspex, Mycalex—a mica based compound with good thermal and mechanical properties and capable of being machined—and epoxy based materials.

The electrical properties of these materials are given below.

Material	Dielectric Constant	Puncture Voltage*
Air (dry)	1.0	~22
Bakelite/Hylam	4.0	700
Glass	8.0	200
Mica	7.0	1500
Mycalex	7.4	250
Paper	2.0	1250
Polystyrene	2.9	2500
Porcelain	7.5	100
Rubber	3.5	450
Teflon	1.9	700
*In V per mil (0.001 inch)		

Epoxy Resins

Epoxy resin formulations are important in the electronics industry, and are employed in motors, generators, transformers, switchgear, bushings and insulators. Epoxy resins are excellent electrical insulators and protect electrical components from short circuiting, dust and moisture. In the electronics industry, epoxy resins are the primary resin used in over-moulding integrated circuits and hybrid circuits, and making printed circuit boards. The largest volume type of circuit board—an "FR-4 board"—is a sandwich of layers of glass cloth bonded into a composite by an epoxy resin.

Epoxy resins are used to bond copper foil to circuit board substrates, and are a component of the solder mask on many circuit boards. Flexible epoxy resins are used for potting transformers and inductors. By using vacuum impregnation on uncured epoxy, winding-to-winding, winding-to-core, and winding-to-insulator air voids are eliminated. The cured epoxy is an electrical insulator and a much better conductor of heat than air. Transformer and inductor hot spots are greatly reduced, giving the component a stable and longer life than un-potted product. Epoxy resins are applied using the technology of resin dispensing.

As adhesives, epoxies bond in the following three ways:

(a) Mechanically, as the bonding surfaces are roughened;

(b) By proximity, as the cured resins are physically so close to the bonding surfaces that they are hard to separate; and

(C) Ionically, as the epoxy resins form ionic bonds at an atomic level with the bonding surfaces

The last one is substantially the strongest of the three.

By contrast, polyester resins can only bond using the first two of these, which greatly reduces their utility as adhesives and in marine repair.

Mercury

Mercury is metal, silver-coloured liquid in room temperature, has high density and high surface-tension. It breaks up easily into many small droplets and dissolves some metals giving amalgams. It is slightly soluble in water and is naturally occurring. It expands and contracts evenly with temperature changes, a property which leads it to be widely used in thermometers. Its flammability class is non-combustible liquid.

Physical properties of Mercury

- **Boiling point:** 357 °C, enabling its use in thermometers up to 120 °C.
- **Electrical conductivity:** 0.0104 × 106/cm Ohm. Since this is a good conductor of electricity, it is used to 'wet' electrical contacts in switches, relays, etc.

- **Thermal conductivity:** 0.0834 W/cmK
- **Density:** 13.546 g/cm3 @ 300K
- **Melting point:** -38.72 °C. This property enables its use in thermometers down to –20 °C.

Toxicity of Mercury and its Compounds

Mercury absorption is through lungs, skin and stomach. It is most toxic of all organic forms are methyl or di-methyl mercury. Inorganic compounds can be irritating or corrosive to the skin, eyes and mucus membranes. When swallowed, these compounds can cause nausea, vomiting, diarrhoea or kidney damage.

Acute Health Effect

A very high exposure to mercury vapours in the air may cause acute poisoning. Symptoms usually begin with cough, chest tightness and breathing difficulty. Direct exposure to mercury vapour can cause discolouration of lens in the eye.

Hence take care when handling mercury!

PART III

Mini Projects in Instrumentation

Measurement of Power Consumption by Appliances

21

It is often necessary to know how much current is being consumed by any electrical appliance when it is connected on the main AC line. This will indicate the power rating of the instrument.

This information serves two purposes. Firstly, it will confirm that the power rating stamped on the appliance is correct. Secondly, it will also indicate whether it is safe to use the appliance on the AC line where it will be used.

Take the example of an air conditioner. If the power consumed by the device is 3000 watts, it cannot be used on a 10 amp AC line, since it can handle only 2300 watts; much below that is required by the air conditioner.

Under this project, we develop a watt meter to indicate how much of actual power is being consumed by the device under test.

A simple circuit is shown in Figure 21.1, which measures the power consumption of any load connected directly to the AC main line.

A step down transformer is used in this instrument. The secondary winding of 6 V at 3 amps is used in series with the load to be tested. The original primary winding of 230 V is used as a secondary winding and connected to a dc milliammeter through a diode and a potentiometer.

The voltage generated in the secondary winding is proportional to the current through the 6 V primary, which in turn is proportional to the load under test.

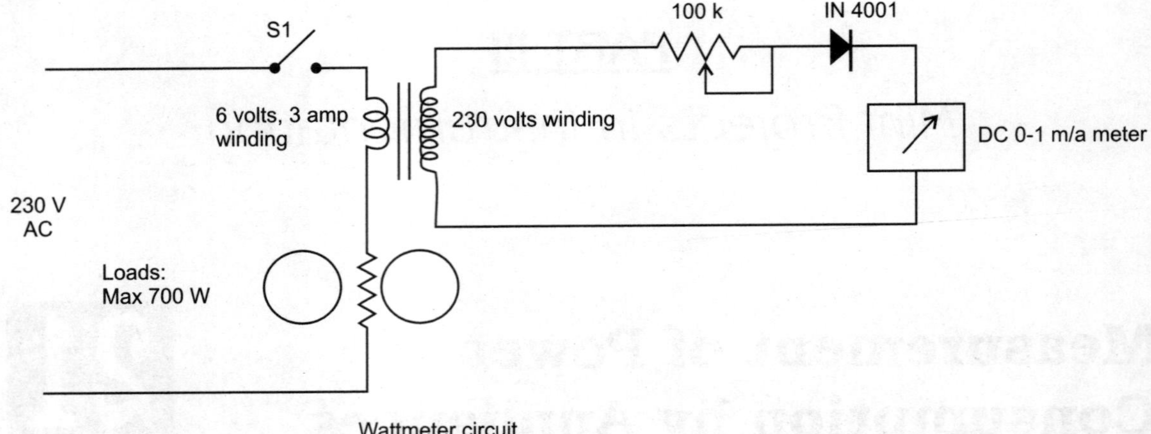

Figure 21.1 *Measuring power consumption of AC loads*

The secondary voltage is rectified by the diode in series with the DC milliammeter. The 100k potentiometer serves to calibrate the circuit so that the output current is a direct indication of the load.

To calibrate the circuit, we need a standard load. A filament bulb with a rating of 60-100 watts is all that is required; however, such ratings on the bulbs are not very accurate. This can be checked by using the basic circuit given in Figure 21.2 below.

To determine whether a 100 watt bulb is actually consuming 100 watts, it is hooked up in series with an AC ammeter which can read up to 1 amp. By using the formula $W = V \times I$, the actual rating of the bulb can be determined. For a bulb of 100 watts rating, we should obtain a current of 100/230 or 0.435 amps (435 milliamps). An input voltage of 230 V is assumed here.

A necessary correction is required if the voltage under test is of a different value.

It should be kept in mind that when the bulb is switched on, there will be a slight surge in current since the filament will be cool, and will decrease thereafter to a steady value when the bulb is fully lit up.

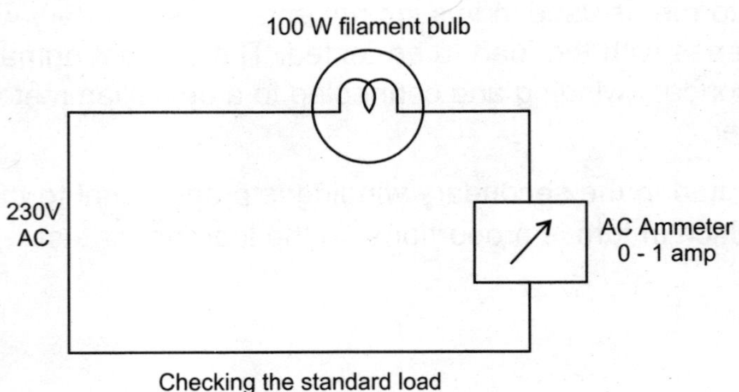

Figure 21.2 *Use a filament bulb as a standard load.*

Once the power rating of the bulb is ascertained, the same bulb can be used to calibrate the circuit, as shown in Figure 21.1.

By using different loads (calibrated as above), a curve can be plotted from which the actual load of any device can be determined.

If loads of more than 700 watts are to be checked, a transformer with a primary rating of more than 3 amps should be used.

The potentiometer, once set with a standard load, should not be altered subsequently.

Any other type of load can serve as a standard; a filament bulb is suggested here due to its easy availability as well as the wide choice of power ratings—commonly from 15 watts up to 200 watts.

Wire-wound resistances can be used in place of the bulbs. They must have the necessary power rating. They are also susceptible to changes in their resistance value, when they get heated.

22 Anemometer—Measuring the Speed of Wind

An anemometer is an instrument which measures the speed of wind or of any other gas, for that matter. It is a device that finds important applications in meteorology, as well as a handy instrument for hobby sailors. The basic instrument has two, three or four hemispherical cups mounted at the ends of radial arms, assembled so as to rotate in a horizontal plane. The higher the wind speed, the faster is the rotation of radial-arm assembly.

The complete instrument has two parts. The first is the radial arm assembly and the second is a device which converts the rotational movement of the radial arm into an electrical signal, which finally indicates the wind speed.

In practice, the radial arm which carries two hemispherical cups has a problem—if the cups are not aligned directly into the wind flow, there will be no rotation. Three hemispherical cups overcome this problem while four cups are the best option.

Figure 22.1 below shows features of the main assembly. The hemispherical cups can be made by cutting a hollow sphere (of metal or plastic) in half. Two such spheres will provide the four cups for the anemometer. The cups are fitted on to a radial arm made up from two metallic strips of the dimensions. The strips are fixed at right angles to each other and bolted at the centre. Inset 1 in Figure 22.1 shows how the cross strips fit over each other at the centre by making a slot in each of the arms and fixing the strips so that they are at the same horizontal level after being bolted.

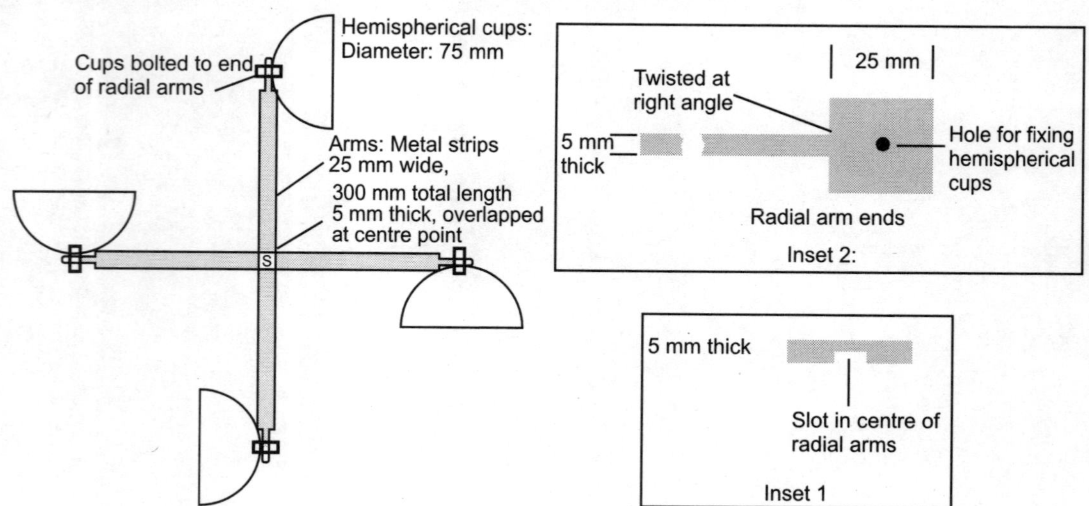

Figure 22.1 *Anemometer main assembly with hemispherical cups*

The revolving radial arm rotates the armature of DC motor, through a coupling, as shown in Figure 22.2. This is shown in coloured illustrations on pages 177–184.

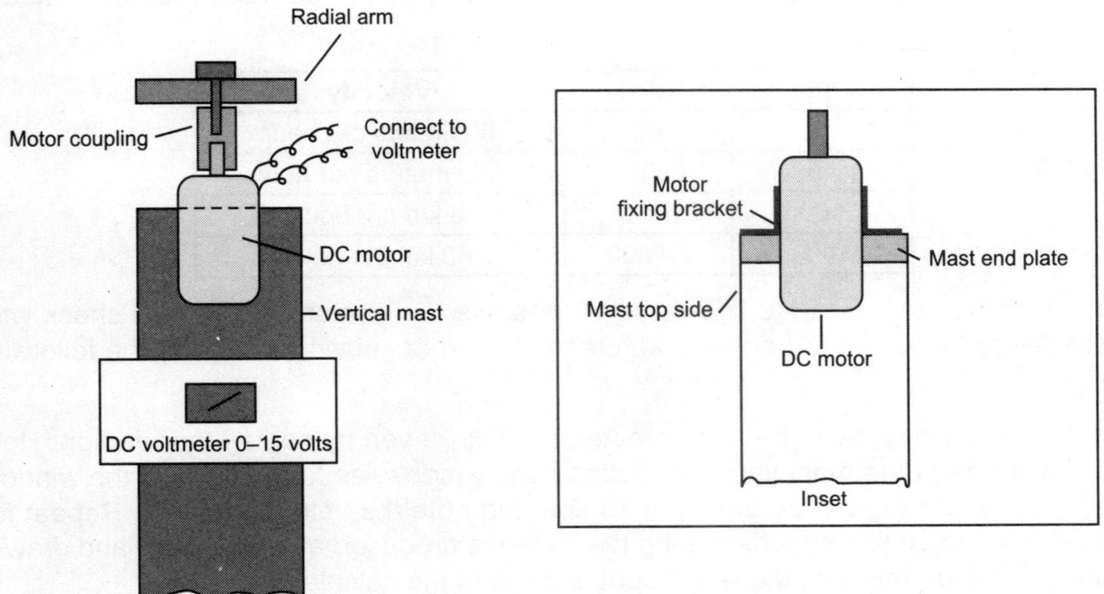

Figure 22.2 *Assembly of motor and voltmeter on mast*

The DC motor acts like a dynamo and generates a voltage which is linearly dependant on the rotational speed of its armature. The voltage output from the motor is then an indication of the motor speed.

But how do we determine the wind velocity? To do this, we must determine the RPM versus voltage characteristics of the motor. This can be done by using another motor whose speed is known at a specific voltage.

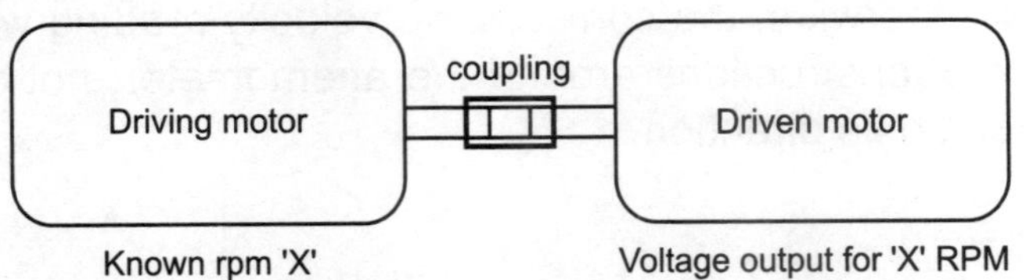

Figure 22.3 *Determining the voltage output characteristic of a DC motor*

We need to know the voltage output of the DC motor at least at two different speeds, so as to draw a graph of RPM versus voltage output.

Once this is done, we can determine the RPM from the voltage generated by the motor at different wind velocities. But what about the wind velocity?

A simple calculation shows how this can be done.

The distance travelled by the hemispheres in each revolution will be equal to its circumference. If the mean distance between the hemispheres is 318 mm, the circumference of the circle traced by the hemispheres will be 1000 mm, or 1 meter. Then—

RPM	RPH	Velocity
1	60	60 meters per hour
10	600	600 meters per hour
100	6000	6 km per hour
1000	60,000	60 km per hour

To determine how accurate this calculation is, we can perform a physical check with known wind velocities. To do this, we create a wind of known velocity by the following method.

On a windless day, take the anemometer in a car, driven by some other person. Hold the lower end of the mast vertically outside one window, as far away from the window as possible. Let the car be driven at 10 kph and note the voltage reading. Repeat for 20, 30, 40 and 50 kph speeds, noting the voltages produced at each speed and draw a curve. Compare this with the results obtained from the calculations.

To perform this experiment in a moving car, the vertical mast must be of a length which is convenient to hold (below the voltmeter); hence, the voltmeter must be as near to the top of the mast as possible. The wires from the motor are directly connected to the voltmeter here.

However, if the anemometer is to be fixed on to the ground, then the mast should be at least 5 meters tall, over the ground. A suitable grounding arrangement has to be made so that the mast does not topple in high wind velocities.

It is also necessary that the anemometer is not surrounded by buildings and has a clear access to the winds; otherwise, the correct wind velocity reading will not be obtained. Due to the presence of obstructions around the anemometer, not only does the wind velocity change, but also its direction.

Logic Status Determination Through Audible Signals

23

Determination of logic levels at various points in digital circuits is a routine exercise. Different kinds of logic probes are available with varying degrees of complexity. Most of these give some kind of a visual output, either through LEDs or even seven segment displays. Displays on computer monitors are also current practice, especially if simultaneous display of logic levels at multiple points is required.

One drawback of such displays is that while checking the logic level at any particular point, observation of the location as well as the readout has to be done simultaneously. To overcome this drawback, audible signals can be generated with differing tones to signify different logic levels. One such logic probe is described under this project.

The electronic circuit is simple, requiring only one TTL IC, two transistors and a few passive components. A power supply to provide the circuit with DC power is also required. The various parts of the instrument are described below.

In digital circuits, the signals are composed of a string of 0s and 1s. Thus, the instrument has to provide information as to which of these two states exist at the point, where the test is being carried out. Figure 23.1 below shows the high and low levels in a digital signal.

Digital circuits now form an essential component of almost all tests, measuring and analytical instruments. Such circuits are based on sub-systems, which too are digital in nature and grouped together to form the complete system. Digital computer is a good example of such systems.

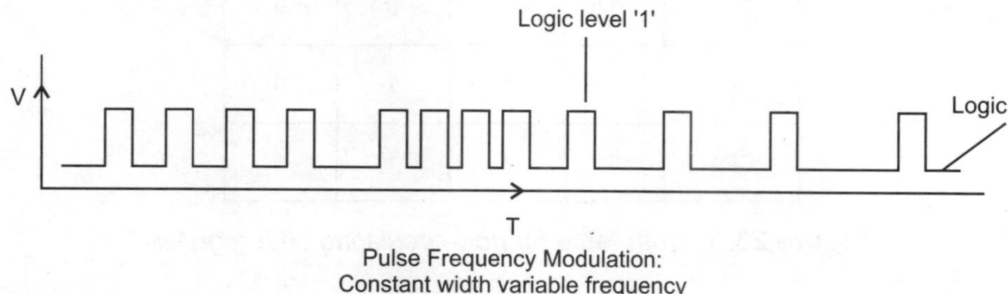

Figure 23.1 *Logic levels in a digital signal*

To explain the nature of digital circuits, an example is given in Figure 23.2, which is a four stage non-circulating shift register, based on TTL ICs. Also given is the 'Truth

Table' for the same circuit (Figure 23.3), which indicates the logic levels at various stages of the shift register.

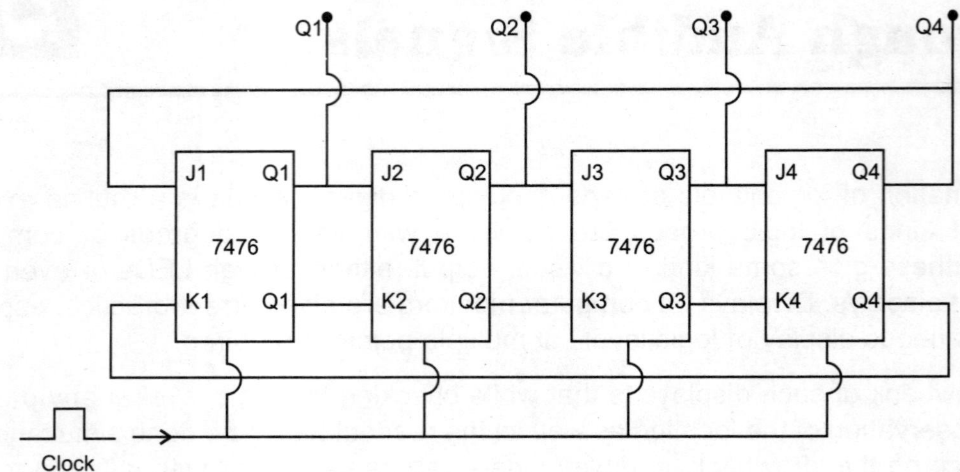

Figure 23.2 *Four stage non-circulating shift register*

As may be observed from the Truth Table, the high and low levels keep changing with each incoming clock pulse. A logic probe is useful here to ascertain that the shift register is working satisfactorily.

In this particular example, the logic probe will confirm that all outputs finally end up at high level, with clock pulse 4.

Truth Table	Q1	Q2	Q3	Q4
RST	0	0	0	0
CP1	1	0	0	0
CP2	1	1	0	0
CP3	1	1	1	0
CP4	1	1	1	1

Figure 23.3 *Truth table for non-circulating shift register*

The Logic Probe with Audio Output

The complete instrument is composed of two distinct units:

(a) The logic probe with built-in electronic circuit.

(b) A power supply for the probe unit.

These are shown in Figure 23.4 below.

The probe circuitry is shown in Figure 23.4. The circuit is built around a single IC, 7402, comprising four NOR gates, connected in two pairs, to form two oscillators running at two different frequencies (in the audible range), which are dependent on their RC networks.

Thus, the oscillator section at the bottom runs at ten times the frequency of the top section.

When the probe contacts a high level, the input to the NOR gate goes high and the oscillator operates at a lower frequency. When the probe contacts a low level, the top oscillator cannot function, while the lower one is switched on due to the inversion by the transistor BC107. The output is then a high pitched sound.

The two diodes serve to isolate two oscillators. The audio output is provided by a piezo speaker.

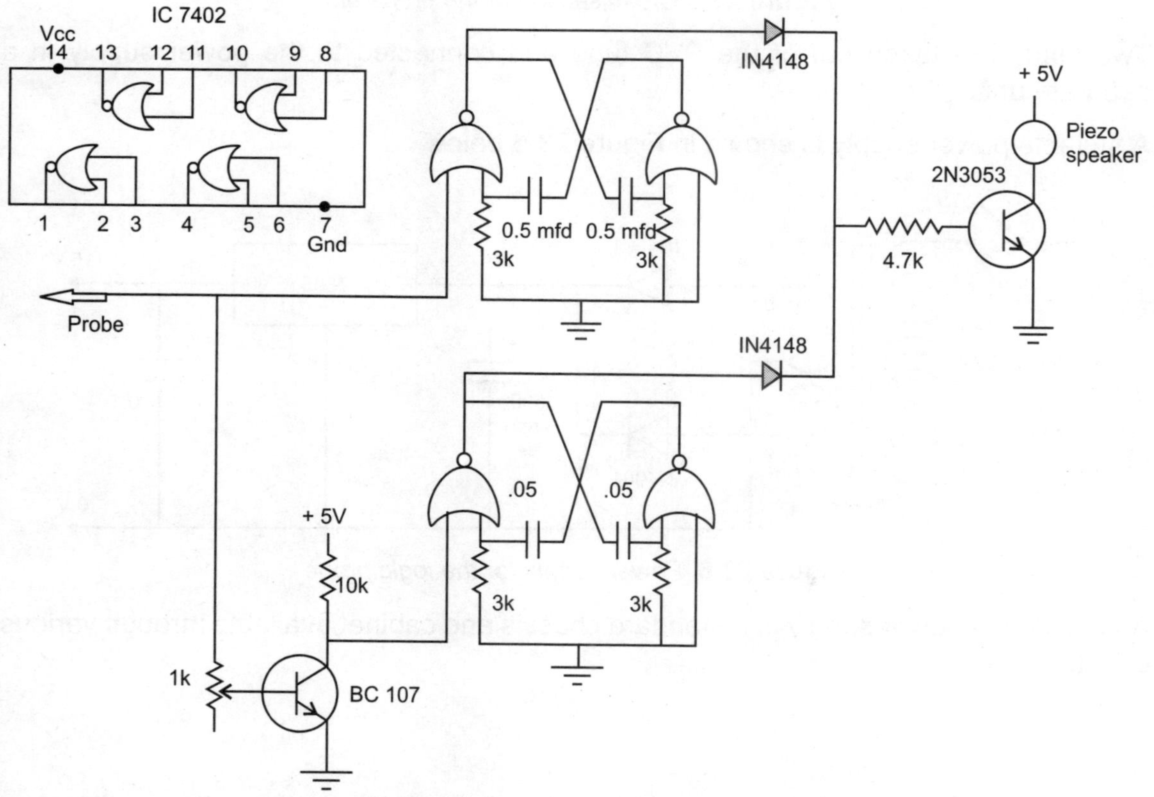

Figure 23.4 *Circuit diagram of probe electronics*

The two diodes serve to isolate the two oscillators. The audio output is provided by a piezo speaker.

The circuit is assembled on a small perfboard, so that it can be accommodated inside the probe housing, as shown in Figure 23.5.

The probe and circuit board are housed inside a PVC tube. One end of this carries the probe head. This can be part of a discarded meter test prod, fixing into an insulator, as shown in the figure above. The other end of the PVC tube accommodates a piezo speaker of appropriate size.

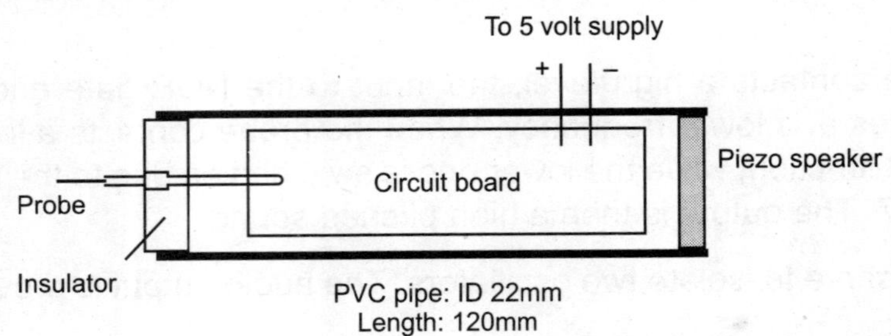

Figure 23.5 *Cross section of the probe unit*

Two leads are taken out of the PVC tube and connected to the power supply in a separate unit.

A suitable power supply is shown in Figure 23.6 below.

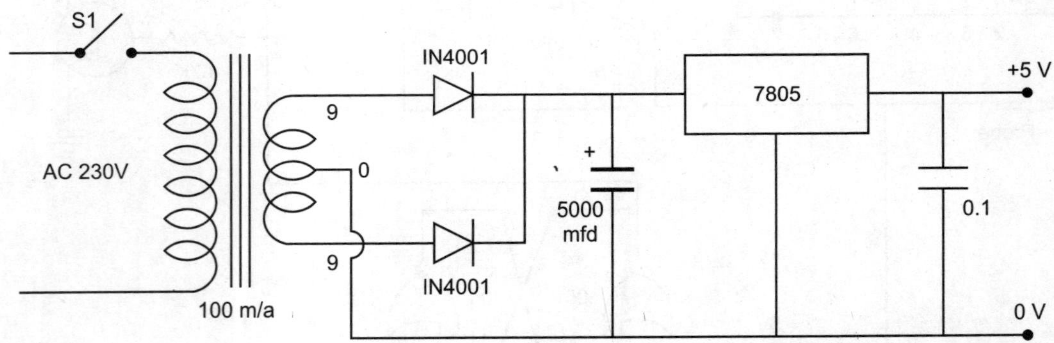

Figure 23.6 *Power supply for the logic probe*

Assemble the power supply in a standard chassis and cabinet available through various outlets.

Measure Weak Currents with Self-made Instrument

24

The most common types of instrument for measuring DC voltages and currents are the moving coil instruments, which are based on the interaction of an electric field with a magnetic one. In these instruments, the magnetic field is produced by a stationary permanent magnet, while the current to be measured is fed to a coil suspended between the magnetic poles. The movement of the coil within this area is proportional to the current through the coil. A needle fixed on the moving coil moves over a calibrated scale, which indicates the current through the meter.

A galvanometer, of which there are different varieties, is also an instrument to measure direct currents. Galvanometers may be divided broadly into two types—the moving magnet and the moving coil. For alternating currents, a vibration galvanometer can be used, in which a string vibrates in synchronism with the AC through it.

The behaviour of a galvanometer can be described by its working constants. These are electric resistance, sensitivity, free period and critical damping resistance. Its equation of motion is given as:

$$P\frac{d^2\theta}{dt^2} + \left(K + \frac{G^2}{R}\right)\frac{d\theta}{dt} + U\theta = \frac{GE}{R}$$

Where P is the moment of inertia; K is the mechanical damping coefficient; G is the motor constant; R is the circuit resistance and U is the restoring torque constant.

A moving magnet type of galvanometer can be easily fabricated and is described in this project. It is assembled using the following hardware:

(a) PVC pipe to serve as the coil former. This is selected from one of the standard types available in hardware stores. The diameter is 132 mm and a length of 25 mm is required.

(b) The end pieces for the coil former are fabricated from 6 mm plywood of 180 mm diameter.

(c) Enamelled copper wire of 18 SWG. A length of 8 meters is required.

(d) Circular mirror 25 mm in diameter—see fabrication techniques.

(e) Key chain, 60 mm length.

(f) Terminal for connecting wires—2 numbers.

(g) A small bar magnet of the type available in toy stores. Maximum length is 20 mm.

(h) Base plate with spacers.

(i) Laser pointer.

(j) Stand for laser pointer.

(k) A transparent acrylic strip 50 mm wide and 150 mm long, with calibrations.

(l) Assorted hardware.

The accompanying figures illustrate various hardware mentioned above.

Fabricating the Galvanometer

Coil assembly

The former on which the coil is wound is a section of a PVC pipe of 132 mm diameter. This is a common size and is available in hardware stores. After a length of 25 mm has been cut, the ends are filed to ensure that they are smooth. A fine hole of 2mm diameter is drilled in the centre of this former. See Figure 24.1. This hole will accommodate a wire hook from which the key chain is hung and magnet-mirror disc is fixed.

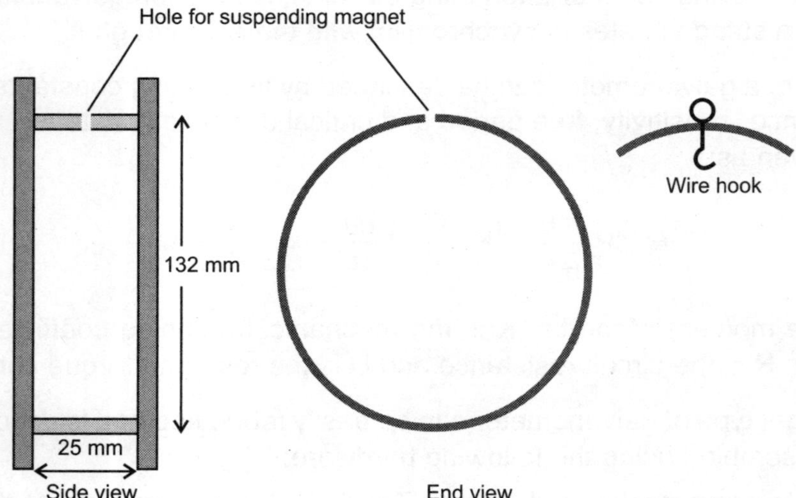

Figure 24.1 *Former with hole for suspending magnet with a wire hook*

Two circular pieces of 6 mm plywood are fabricated. These will act as the ends of the former and will hold the coil in position. In one of the circular discs, a hole of 50 mm is made in the centre. This will be the front side of the coil assembly and the mirror will be visible through this hole.

The plywood discs are glued to the PVC pipe symmetrically. A wire hook is made and fixed in position. The coil is now wound on the former with the 18 gauge enamelled copper wire. The winding should be evenly spread on the former, on either side of the wire hook. Wind twenty turns and leaves enough wire at the ends to connect to the terminals. The lead wires should be brought out from the lower side of the former through two holes made in the rear plywood disc.

Figure 24.2B shows the coil wound on the former. In Figure 24.2A, the mirror is shown suspended in the coil. The coil former is fixed to the base plate by means of two L brackets.

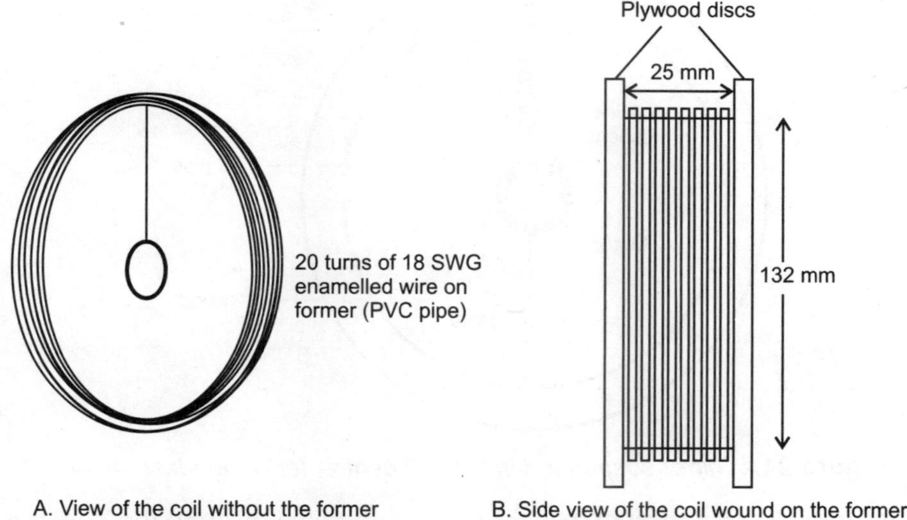

Figure 24.2 *The coil shown with and without the former*

Mirror

A circular disc of Bakelite/Hylam 2 mm thick and 30 mm diameter is prepared first. A hole of 2 mm is drilled on one edge to accommodate the suspension chain. This forms the back plate for the mirror.

The mirror is made from reflective aluminium foils available in hardware shops. A circular piece of 20 mm is cut and pasted on one side of the back plate. See this Figure 24.3 in coloured illustrations on pages 177–184.

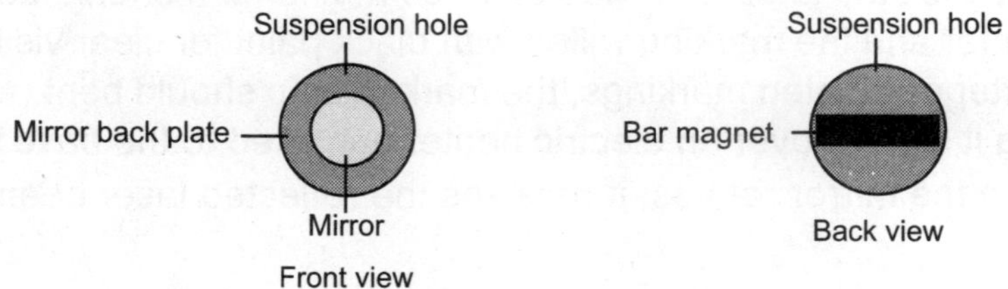

Figure 24.3 *Fabrication of the mirror-magnet unit*

The bar magnet is stuck horizontally on the rear side of the mirror on the back plate. Use a suitable epoxy resin for this purpose.

Mirror Suspension

The mirror is suspended within the coil by means of a key chain, hung from the coil former. The length of the chain should be selected in a way that the mirror is located centrally within the coil. See Figure 24.4 for better understanding.

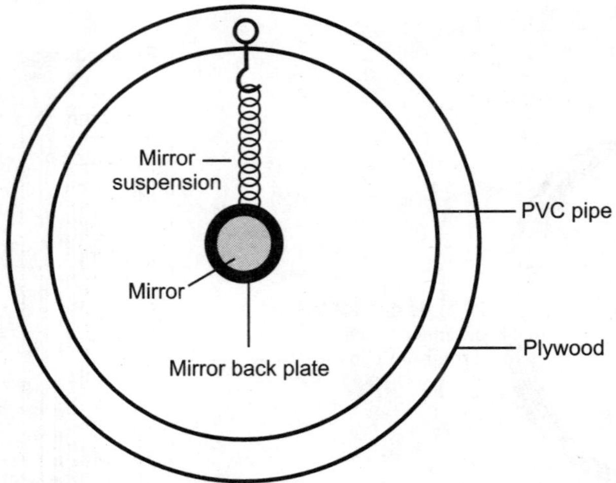

Figure 24.4 *Mirror suspended in the coil centre, facing the laser beam*

The mirror rotates along a vertical axis when current flows through the coil, the rotation being proportional to the magnitude of current in the circuit. When there is no current flow, the plane of the mirror should be parallel to the coil former discs. The front disc has a circular hole, which has an opening. This permits the laser beam to impinge upon the mirror and get reflected to the calibrated acrylic strip located close to the laser.

Laser

A hand held laser pointer, available in stationery stores is used as a source of light. This is fixed directly in front of the mirror at some distance away by means of a stand which should be fabricated to suit the mechanical size of the laser.

The calibrated acrylic strip is 50 mm wide and 150 mm long. It should be marked using a sharp steel pointer and the marking willed with black paint for clear visibility as shown in Figure 24.5. After graduated markings, the marked strip should bent (as shown in the figure) by heating it slightly over an electric heater and fixed to the base plate in such a position that when the mirror rotates, it receives the reflected laser beam.

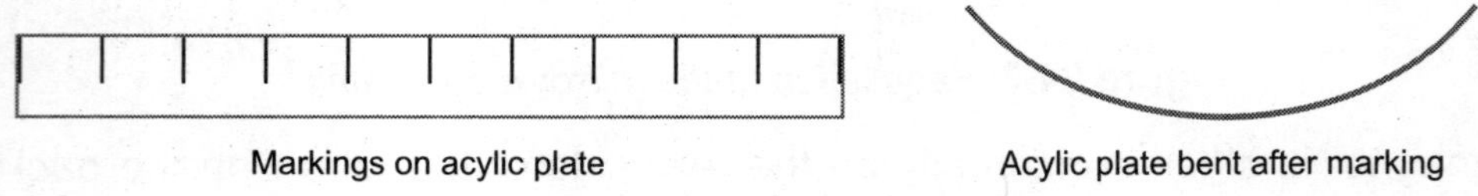

Figure 24.5 *The acrylic strip marked and then bent*

This completes assembly of the galvanometer. The complete assembly is shown in Figure 24.6 in coloured illustrations on pages 177–184.

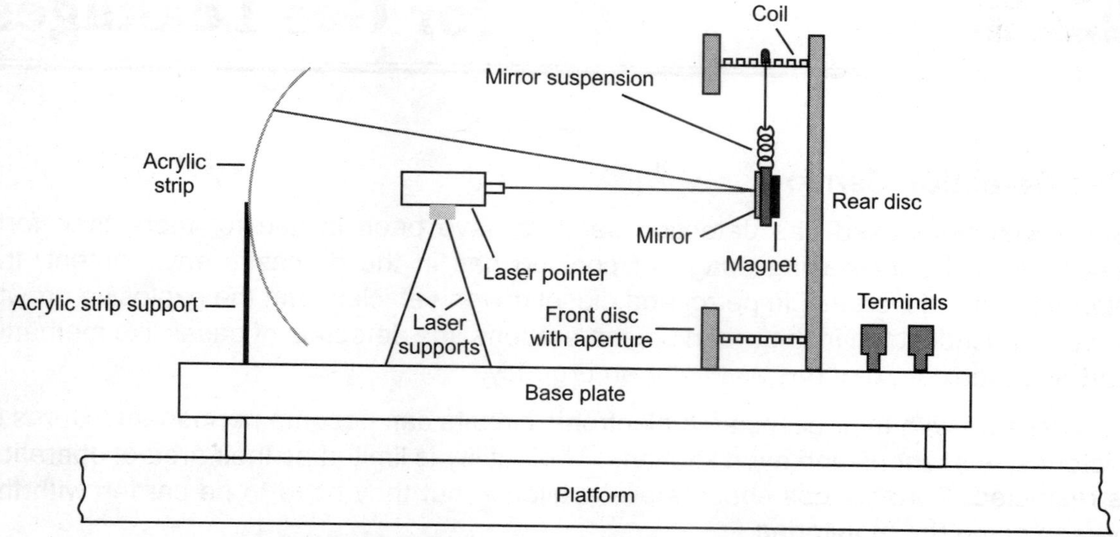

Figure 24.6 *Overall view of the galvanometer assembly*

Two terminals fixed on to the base plate enable the connection to the coil and the current source.

Testing and Application

With the coil connected to the two terminals, the galvanometer can now be tested. The coil resistance will be 1 ohm. The laser should be switched on and placed so that the beam hits the mirror at right angle. Use a resistance of 3 ohms in series with the coil and connect a 2 volt battery to the galvanometer. Since the total resistance is now 4 ohms, a current of 500 mA should flow into the galvanometer, and the mirror should turn deflecting the beam horizontally. If the acrylic strip is correctly placed, the beam should hit on the graduated scale. The farther acrylic strip is from the mirror, the greater will be the deflection of the beam. If the mirror turns away from the acrylic strip, insert the battery connections. With different resistances in series with the galvanometer, different readings on the scale will be obtained. An ammeter connected in series with the battery can be used to convert scale readings into the current readings.

25 Environmental Monitoring for Gas Leakages

Gas Detection Sensors

Semiconductor based gas detection sensors have been in use for more than forty years. With the increasing usage of cooking gas in the domestic environment, the alarming rate of increase in petrol and diesel driven vehicles, and the explosive growth of polluting industries in the urban agglomerations, the detection of gases like methane, carbon monoxide, etc. has assumed an urgency.

Gas sensors with their associated electronic circuits can become permanent fixtures in factories, workshops and even kitchens. Their utility is limited as their area of operation is restricted. Portable gas sensors are available, but they have to be carried with the person doing the monitoring.

In this project, we present an instrument which has the following features:

(a) A gas sensor with associated electronic circuitry and an alarm

(b) Use of lazy tong device which extends the sensor as desired, within its range of operation

(c) A revolving mechanism which turns the sensor over a 180° angle

(d) Both operations through radio control, the operator can be a safe distance away from the toxic area.

Gas Sensor

The gas sensor consists of a semiconductor material surrounding a filament heater. The sensor is within a stainless steel body, the front portion of which is a cup-shaped wire mesh of the same material. This allows the gas to flow in and reach the sensitive material, which is Tin Oxide, with chemical formula SnO_2. The chemical, SnO_2, is doped and is n-type semiconductor.

See Figure 25.1 to have a clear understanding of the internal structure of a gas sensor.

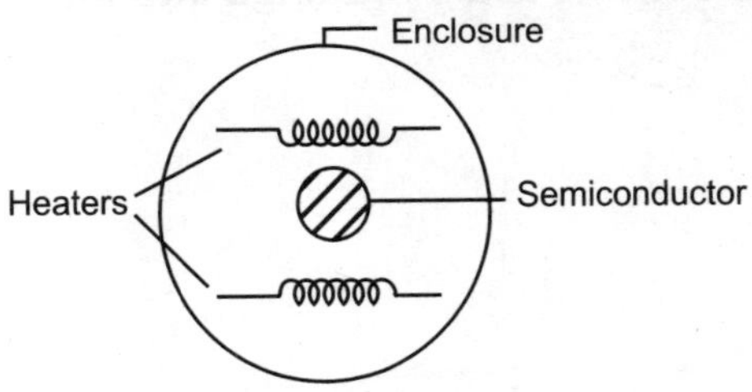

Figure 25.1 *Internal structure of a gas sensor*

The filament heater, when powered, heats the chemical material whose resistance decreases in the presence of certain chemical vapours, as they pass through the mesh and reach tin oxide. Carbon monoxide, hydrogen, alcohol and other combustible gases and vapours have the effect of decreasing the resistance of the tin oxide semiconductor leading to a multitude of applications. Since it can detect the presence of carbon monoxide, it can be used as a smoke detector also.

The semiconductor material is very sensitive to changes in gas concentrations and sense 50–100 ppm (parts per million) of carbon monoxide, etc. The changes, however, are not permanent. They can be reversed enabling the sensor to be used thousands of times during its useful life. However, the sensor cannot distinguish between the various vapours.

In practice, the sensor needs to warm up after being switched on. It will take a few minutes for it to stabilise before it can be used.

Electronics Detection Circuit

An electronics circuit for use with the gas sensor is shown in Figure 25.2. The circuit is powered by a re-chargeable battery of 6 V. The gas sensor, the 555 IC, the Sonalert alarm and other components are assembled on a perfboard with the power cables running through the lazy tong pipes to the main battery container. They terminate here through the switch and connect to the re-chargeable battery.

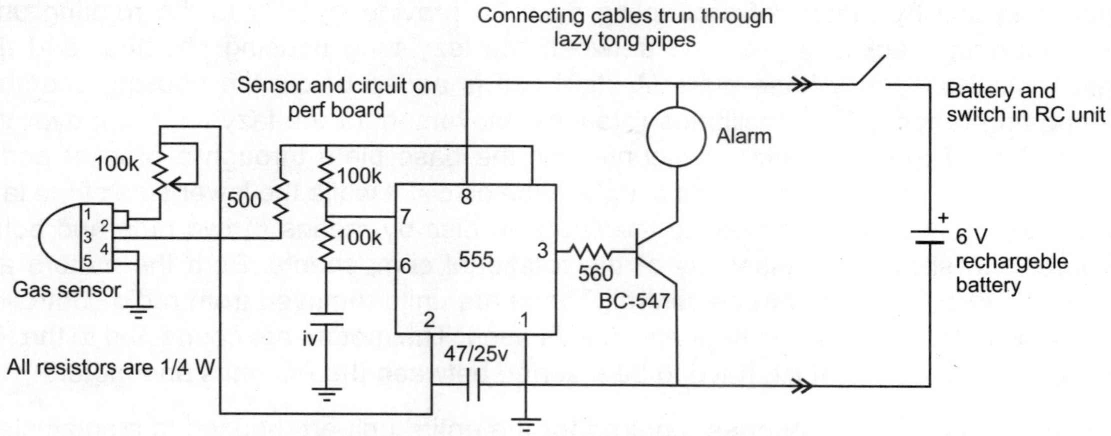

Figure 25.2 *Electronics circuit incorporating a small alarm*

Remote Handling Unit

The mechanical handling unit consists of two major components—the lazy tong device and the turntable. The lazy tong device enables the operator to extend or withdraw the sensor in a horizontal direction. The turntable rotates entire mechanical unit over an angle of 180°. With these two operations, the sensor can be pointed in any direction and over a certain range of distances.

The Lazy Tong

The extension of lazy tong device depends on the number of individual elements that form the structure. The principle of the lazy tong and details of its fabrication are given separately in Part I of this book in Chapter 4 titled Lazy Tong.

In this project, the motor for driving the Lazy Tong is a DC geared motor operating at 6 V. This is a reversible motor to extend the lazy tong or withdraw it within its limit of operation. See Figure 25.3 below for a fair understanding.

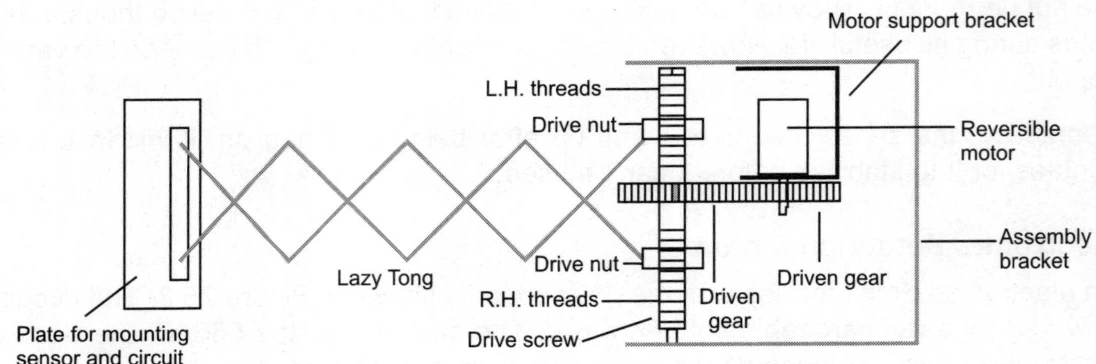

Figure 25.3 *The Lazy Tong unit*

To rotate the lazy tong, an extra motor is required. This is fixed directly below the lazy tong unit by means of a coupling disc. To provide stability to the rotating unit, a supporting bracket is provided between the lazy tong housing chamber and the mechanical assembly base plate. A steel ball is added below the housing chamber supporting bracket. This facilitates rotational movement of the lazy tong unit over the base plate. The drive motor is mounted on the base plate through a bracket and a coupling disc. The motor shaft fits snugly in the disc slot while the lower side of the lazy tong assembly bracket is fixed to the coupling disc by means of two nuts and bolts. Figure 25.5 shows the assembly of the rotational components. Both the motors are operated through radio remote control. These are units removed from radio controlled toy cars and can be used without any modification. The motors are connected to the RC units through relays, which have to be inserted between the RC units and motors.

The RC units and the batteries required for the entire unit are housed in a small steel cabinet directly below the lazy tong unit. The cabinet has to support the weight of the entire lazy tong unit; hence it has to be of appropriate dimensions and strength.

The connecting wires from the gas sensor unit, which come through lazy tong pipes, are connected to the battery through a toggle switch. The connecting wires of the two motors also terminate in the lower chamber.

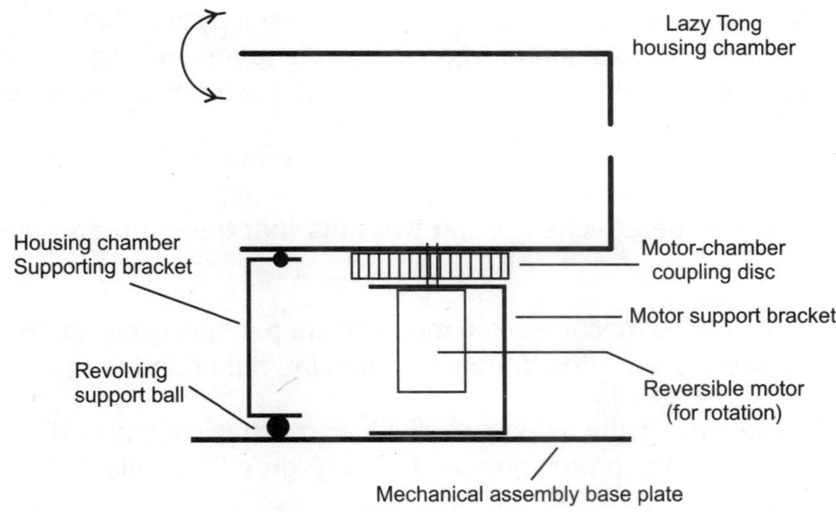

Figure 25.4 *Lazy Tong revolving unit*

The complete assembly of the gas sensor unit is shown in Figure 25.5.

Assembly and Testing

The heart of the instrument is the gas sensor unit. This should be assembled first and then tested. The circuit should be built on a small perfboard, with the sensor having an unobstructed view of the front. Since this circuit is to be mounted on the edge of the lazy tong unit, keep the weight as small as possible—the perfboard should be just large enough to assemble the components. A board of size 50 mm × 75 mm has been found quite adequate for this.

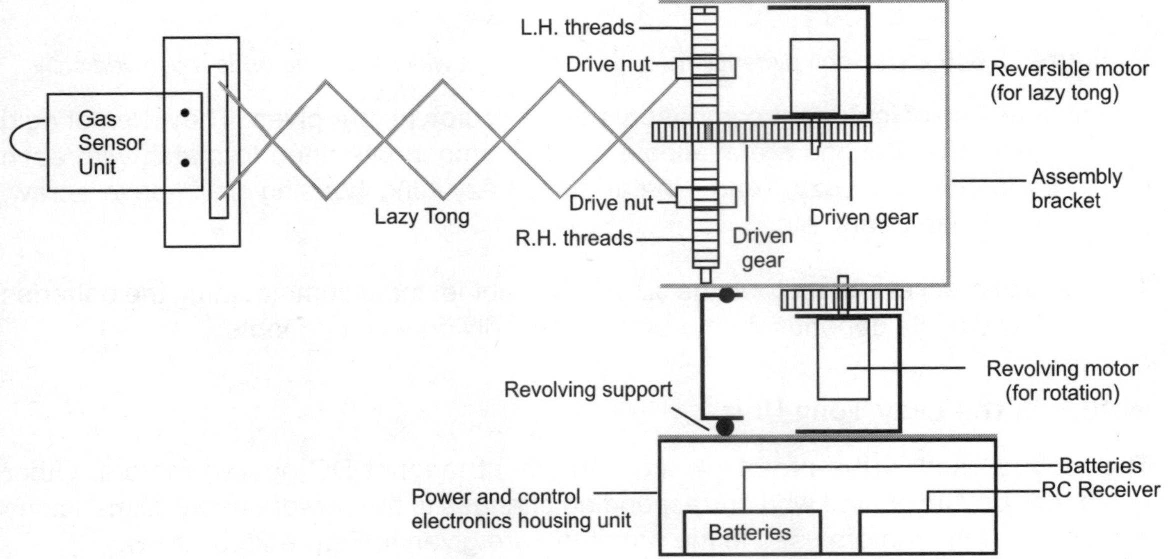

Figure 25.5 *Assembly of gas sensor unit with lazy tong and rotation unit*

After assembly, connect the circuit to a battery and test it by keeping it near a domestic gas cooker. The alarm should sound when the gas is released for a couple of seconds only. Make sure that there are no open flames nearby when this test is carried out.

The lazy tong unit should be built next. The complete mechanical details are provided in Part 1 of this book. Understand how to fabricate the extension arms and the manner in which they have to be attached to the two nuts that travel up and down the drive screw.

The drive screw has to be machined in a machine shop. It has two sections; the top one has left handed threads while the bottom part has the right handed ones.

As a guide to fabricating the various individual mechanical parts, the dimensional drawings of these as in the prototype are given in Figure 25.6 below.

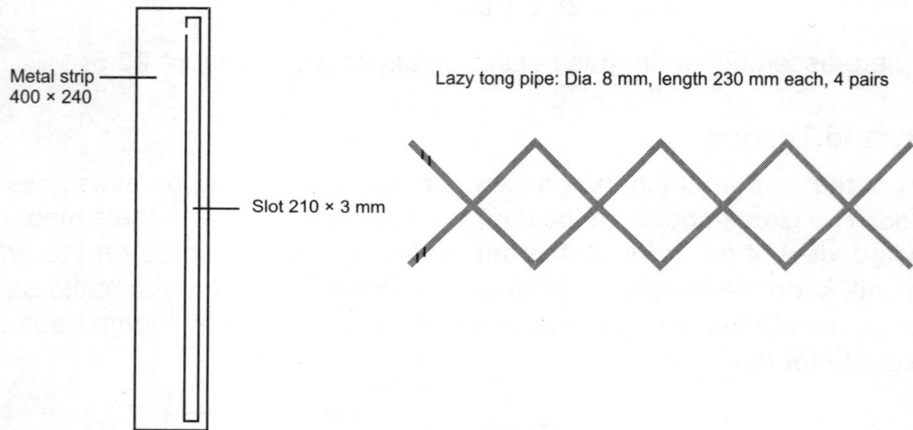

Figure 25.6 *R Lazy Tong dimensions L: Slotted strip in which the tong ends move vertically*

The dimensions of important components of the prototype are given. They are not rigid and various other options are available. Some components need to match with each other. Examples are Lazy Tong—metal strip, Lazy tong housing unit—drive screw, nuts, drive-driven gears, etc.

The sizes of brackets for motors, the size of the cabinet for accommodating the batteries and the RC circuits depends on the size of the individual components.

Motors in the Lazy Tong Unit

The motors used in the prototype are permanent magnet DC geared motors. Other types may be substituted with corresponding changes in the power supply. Dimensions and specs of the motors used in the prototype are given in Figure 25.8 below.

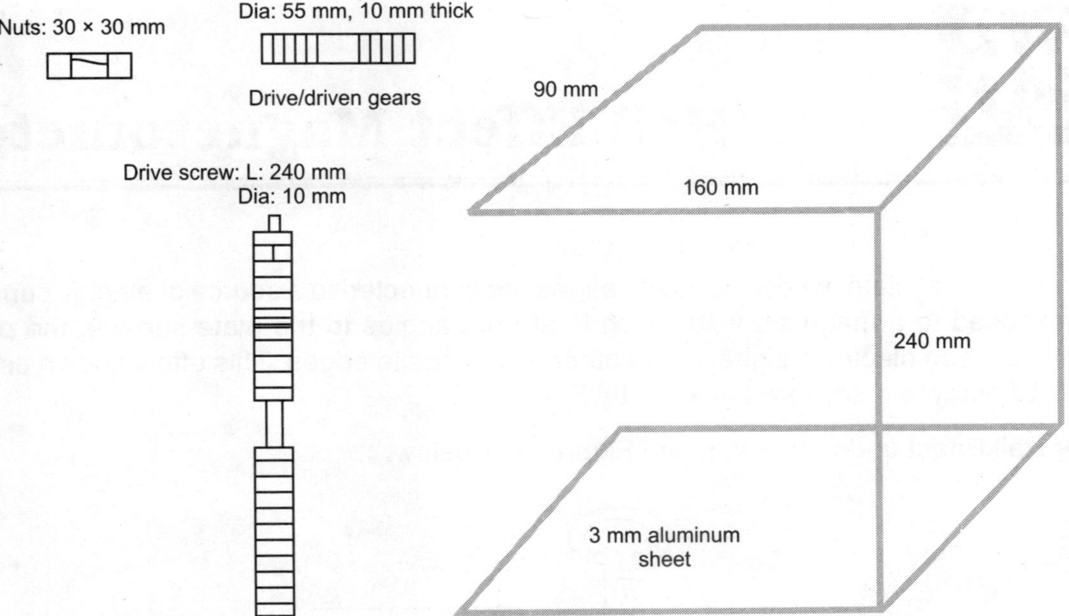

Figure 25.7 *Dimensions of some critical mechanical components*

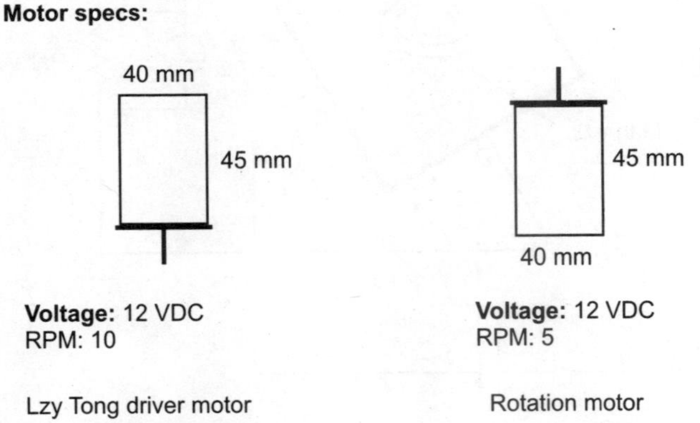

Motor specs:

40 mm

45 mm

Voltage: 12 VDC
RPM: 10

Lzy Tong driver motor

45 mm

40 mm

Voltage: 12 VDC
RPM: 5

Rotation motor

Figure 25.8 *DC motor specifications*

26 Hall Effect Magnetometer

When a metal plate, whose opposite edges are connected to a source of electric current, is exposed to a magnetic field which is at right angles to the plate surface, the plate generates an electrical signal at the other two opposite edges. This effect known as the Hall Effect was discovered in 1879 by E .H. Hall.

For Hall Effect understanding, see Figure 26.1 below.

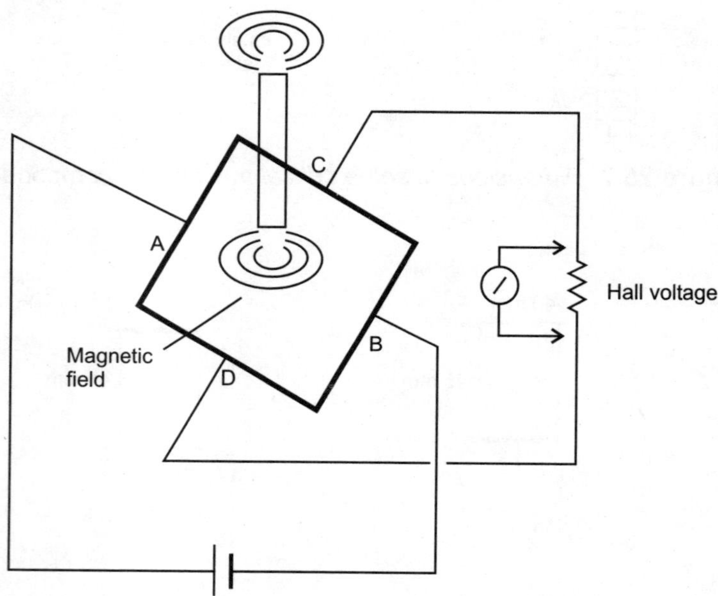

Figure 26.1 *Hall Effect in a metal plate*

Hall Effect is also observed in semiconductor materials like silicon to a large extent. Since the Hall voltage is generated, only when a magnetic field is imposed on the device. It can serve as a sensor for magnetic fields.

The measure of Hall Effect is the Hall Coefficient. It is given by: $R_H = \frac{Ey}{jx\ Hz}$

Where Ey is the electric field developed in the y-direction when a density current jx flows in the x-direction through a magnetic field Hz in the z-direction.

In Hall sensors, the control current is held constant and the resulting Hall voltage depends directly on the magnetic flux density.

Since an electric current flowing through a conductor generates an electric field, Hall sensors can also be used for measuring electric currents on-line. The magnitude of electric current generated in a conductor depends upon the current flowing through the conductor–the more the current, the larger is the magnetic field and correspondingly, the larger will be the Hall voltage.

Hall voltages which are generated in any device are small in magnitude and amplification is necessary for them to be read out on a meter.

A circuit using a Hall sensor to operate a relay is given below in Figure 26.2.

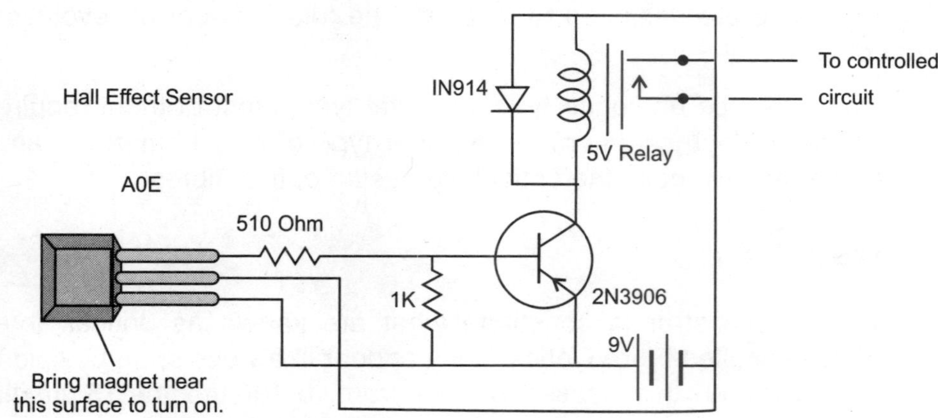

Figure 26.2 *Hall Effect sensor operating a relay*

If a piezo speaker is to be operated directly, the entire assembly can be housed in a square Teflon tube, as shown below in Figure 26.3.

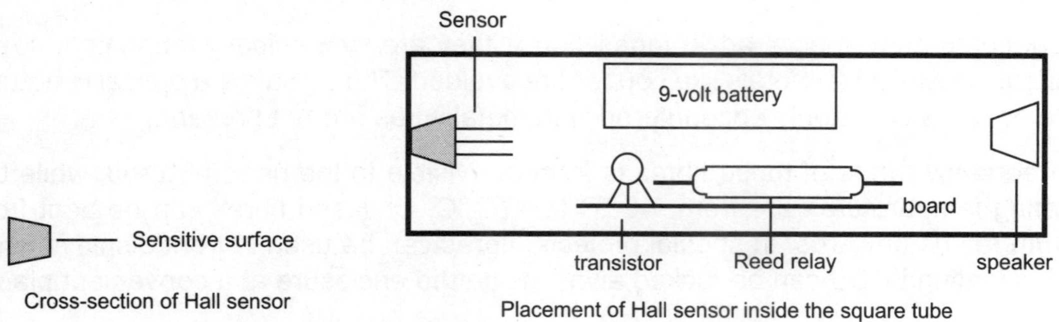

Figure 26.3 *Hall sensor probe with all components inside*

27 Optic FIbre Tachometer

A tachometer is an instrument to measure angular velocity of electric motor shafts or any other rotating object. It is generally used to indicate the number of revolutions per minute, although some are designed to measure the total number of revolutions in a given amount of time.

Tachometers can either be analogue type or digital type. Most of them require some kind of mechanical or electrical interface. Another type of a tachometer is an optical type. Under this project, we build the optical type using optical fibres.

Optical Fibres

Thin, transparent plastic strands constitute what are known as optical fibres; the relevant technology is called Fibre Optics. They appear like webs spun by spiders and are manufactured in various thicknesses, ranging from 10-100 μm; the 15 μm diameter being the most commonly used. Diameters more than 50 μm are stiffer and cannot be bent or twisted readily. Light passes through these fibres by means of multiple total internal reflections, the reflections taking place at the inner side of fibre walls. The efficiency of light transmission is very high and only a very small fraction of the light intensity is absorbed during the reflection process. See Figure 27.1 below.

Optical fibres offer special advantages. Since they are synthetically fabricated, costly electrical transmission metals like copper are avoided. The signals are optical in nature, hence hazards commonly encountered with metal wires are not present.

The frequency range of these fibres is from the visible to the near infra red, while the operating temperatures are from -40 °C to +70 °C. Uncased fibres can be bent from 12.5mm radius onwards. In special projects, fibres can be used in front displays while the illuminating LEDS can be tucked away inside the enclosure at a convenient place.

The principle of the optical fibre tachometer can be understood with reference to Figure 27.1 and 27.2.

There are two optical fibres in this arrangement. One of these is a vibrating fibre and the other is a fixed one. They are arranged in a straight line so that the light from LED passes through both of them, when there is no physical displacement of the vibrating fibre. This light is detected by the photo sensor at the exit end of the stationery fibre.

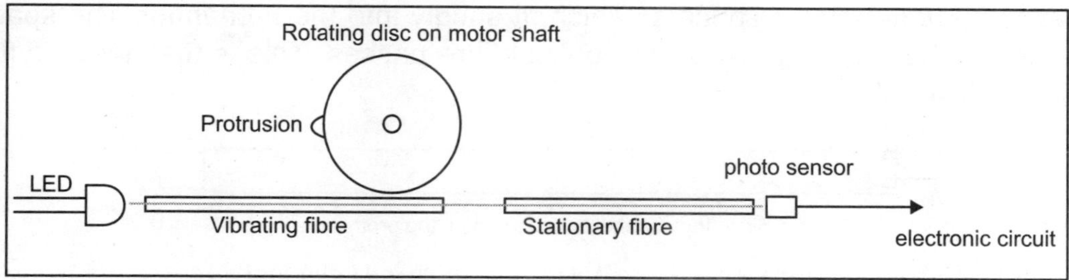

Figure 27.1 *In normal position, the LED ray reaches the photo sensor*

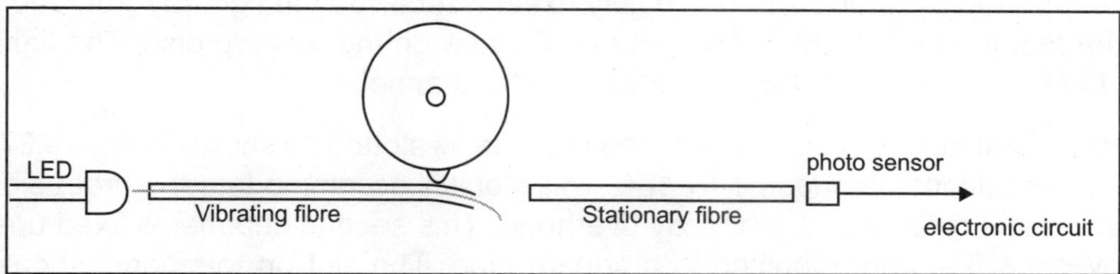

Figure 27.2 *Protrusion on rotating disc causes fibre to bend.*

The rotational speed of the motor shaft can be measured as follows:

In each revolution of the motor shaft, the protrusion on the disc displaces end of the fibre on left side, causing a momentary interruption of the ray of light from led to photo sensor. This results in an electric pulse from the sensor for each revolution of the motor shaft.

The electrical pulses are fed to an electronic circuit, which is designed to convert the frequency of pulses from photo sensor into a current reading, and display this on a milliammeter.

Fabrication of the Fibre Optics Unit

The fibre optics unit is accommodated in two u-shaped aluminium channels of 1.5 cm cross-section each, one on the lower side and the other as a top cover (see Figure 27.3).

Figure 27.3 *Two u-channels fixed upside down from the optical fibre unit*

A red led of 3 mm diameter is fixed at one end of the lower channel. An optical fibre starts from this LED and goes part of the way through this channel. This fibre is supported in two places by non-metallic spacers, which fit snugly into the u-channel. The spacers have a central hole through which the optical fibre passes. This is the vibrating fibre. See Figure 27.4 below.

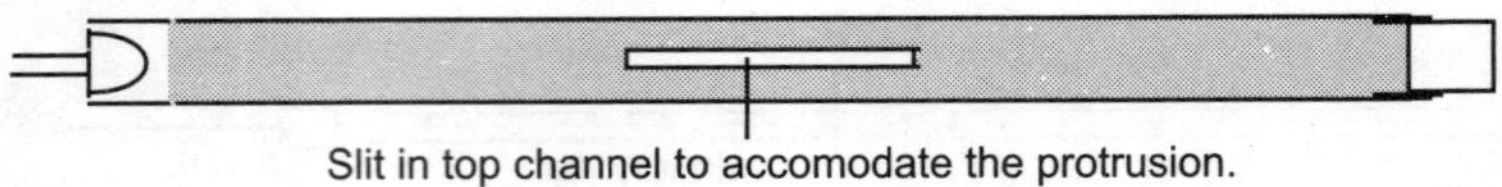

Figure 27.4 *Fibre optics assembly in U channel*

A second optical fibre is fixed after a gap of 2.5 cm, in line with the first fibre. This fibre is also supported by spacers and is rigidly fixed. It receives the light ray from the LED when the first fibre is not deviated by the protrusion on the rotating disc. The light ray strikes the photo sensor at the other end of the u-channel.

A second u-channel, equal in length to the first one, is slotted as shown in Figure 27.5 in coloured illustrations on pages 177–184. This slot will accommodate the protrusion on the disc when it rotates and is directly overhead. This second channel is fixed upside-down over the first one, resulting in a square pipe. The slot on this panel should be directly over the free end of the first fibre.

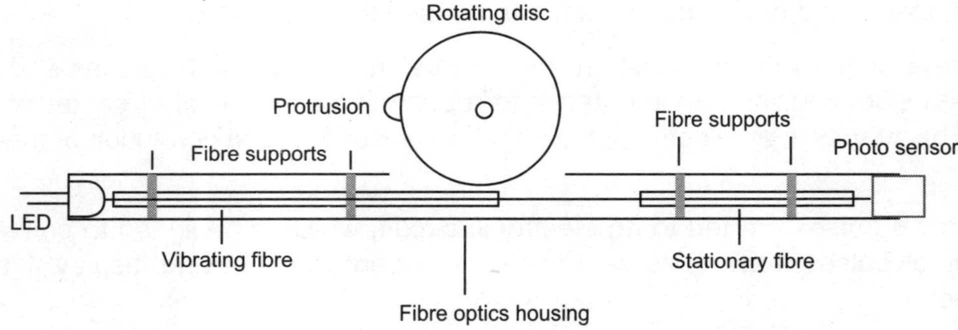

Figure 27.5 *Top channel with slot to accommodate the protrusion on the rotating disc*

When disc rotates, the protrusion passes though the slot and causes the first fibre to bend, thereby diverting the light ray and preventing it to reach photo-sensor. See Figure 27.2 above for clarity.

Electronics Unit

The electronics processing circuit is given is Figure 27.6 below. The input is a phototransistor. As long as phototransistor is receiving the light ray, its collector is at low potential. When the ray is diverted by rotating disc, the collector goes high and then low again, resulting in a positive pulse of 4.5 V for every rotation of the disc.

The pulses from the phototransistor are fed to IC1A, which is configured as a Schmitt Trigger. The amplified signal from pin 1 of this IC goes to 555 after differentiation by

C2-R6 network. The 555 is wired as a monoshot, generating fixed width output pulses at pin 3. These pulses are peak limited by R7 and the 10v Zener. They are then fed to IC1B through the 1N914 diode and the pulse integrator composed of C5 and R9. The voltage generated at pin 5 of IC1B is directly proportional to the frequency of the input signal, which is equal to number of rotations of the motor shaft per minute.

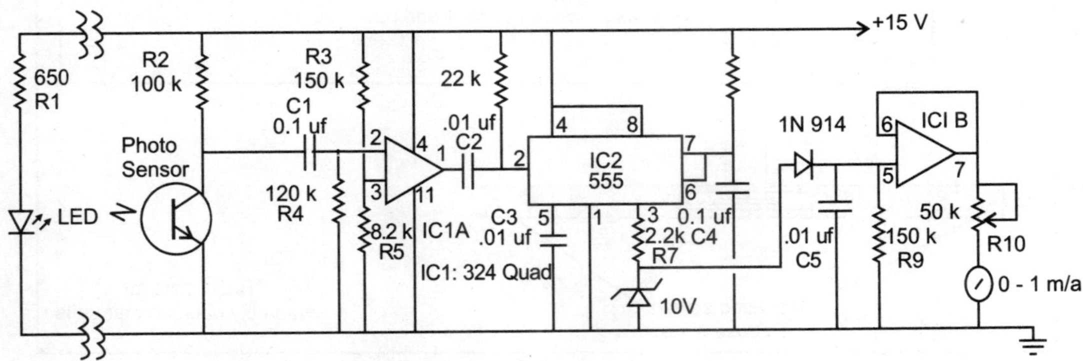

Figure 27.6 *Frequency to voltage convertor circuit*

The potentiometer R10 should be placed to set reading of the meter to full sale (in this case 1 mA), at the highest frequency to be measured (in this case 1 KHz).

Mechanical Assembly of Fibre Optics Components

The lower u-channel is fixed on to a wooden board by means of two screws. The upper channel is then fitted over this. The motor fixture with the mounting bracket and the rotating disc is fixed so that the disc sits into the slot and has a clearance of 1 mm over the fibre optic. When the motor shaft turns, the protrusion on the disc will be upon the optical fibre (as shown in Figure 27.2 above).

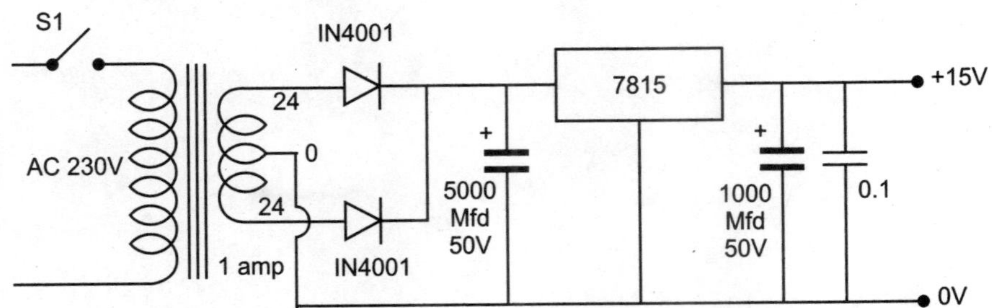

Figure 27.7 *Power supply for the electronics circuit of figure 27.6*

A suitable power supply for the tachometer is given in Figure 27.7 above.

The arrangement of the motor/u-channel unit is shown below in Figure 27.9. The layout of all the sub-units of the tachometer are shown below as mounted on the base plate.

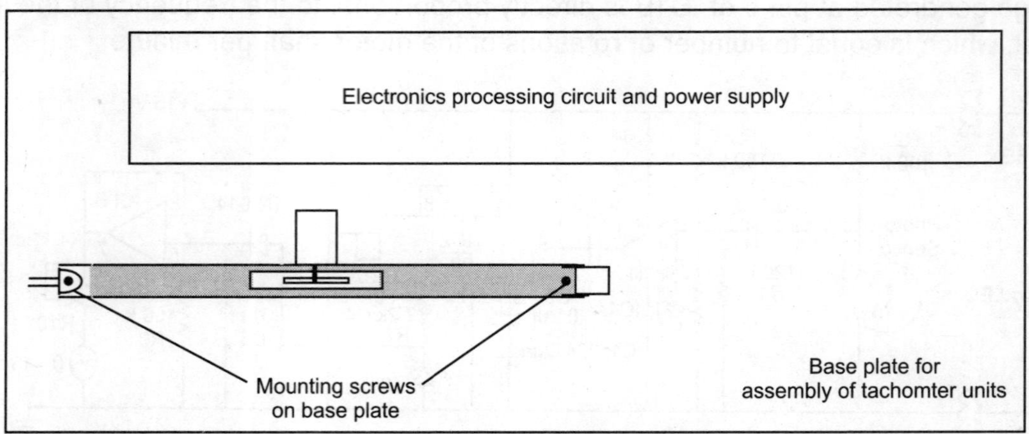

Figure 27.8 *Motor-spindle fixture on base plate over the slot in upper U channel*

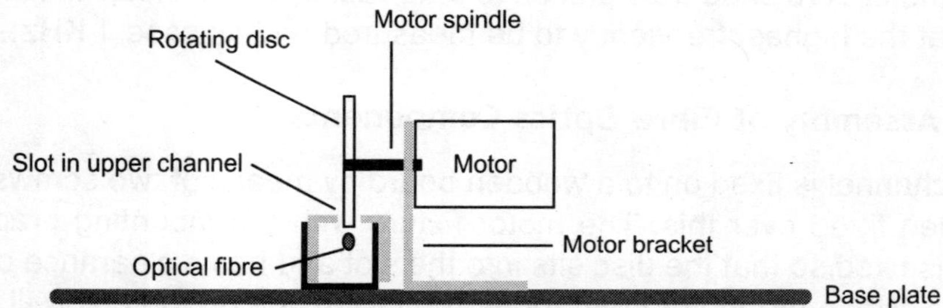

Figure 27.9 *Arrangement of channel assembly and electronic unit*

Instrument to Measure Bullet Velocities

28

Counter chronograph techniques have been described under Chapter 14, Part 1 of this book. Here, we present a project based on this technique to develop an instrument for measuring the velocities of bullets fired from hand guns or air guns.

The basic principles are illustrated in Figure 28.1 and 28.2 below.

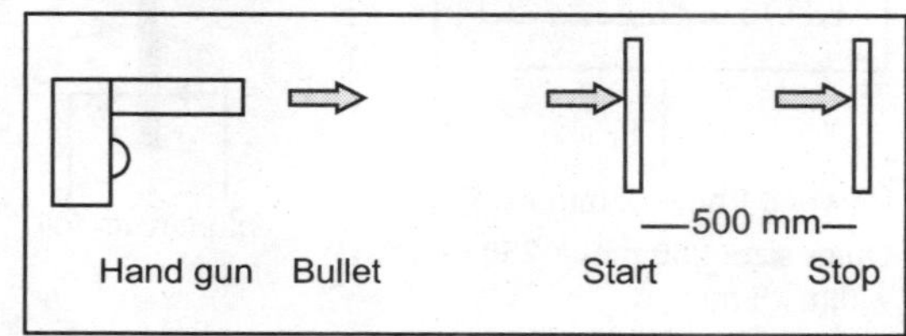

Figure 28.1 *Set up for measuring bullet velocities*

Two sets of mechanical sensors are fixed so that when fired, the bullet goes straight through centre of the sensors. The first sensor starts the time count while second one stops counting of time intervals.

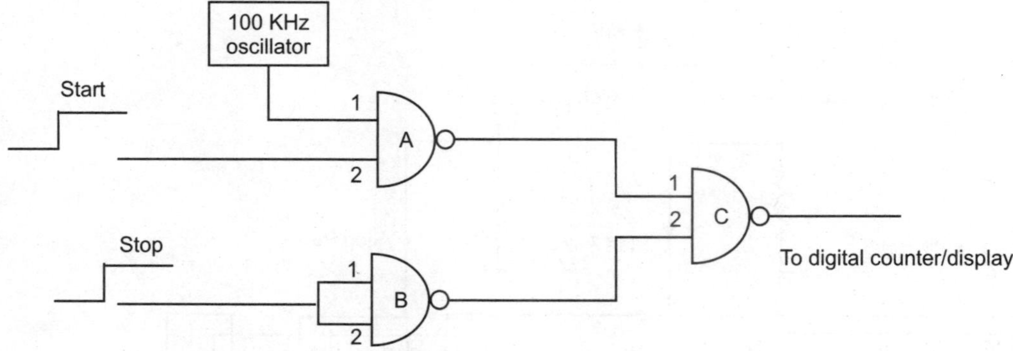

Figure 28.2 *Trigger arrangement for the counter-chronograph*

When the bullet passes through the first sensor, it rips the front foil, which touches the back one, causing it to reach a potential of 5 V. This passes the pulses from the oscillator to the counter/display. Once passing through the second sensor, a potential of 5 V is created, blocking any further pulses from reaching the counter.

The oscillator frequency is fixed at 100 KHz, thus each pulse counted denotes a time period of 10 microseconds. If 200 pulses are counted, then a time period for bullet to traverse the distance of 500 mm is 2000 microseconds. The velocity of the bullet will be 250 m/sec.

The Sensors are shown in Figure 28.3 (1) and (2).

A plywood frame fabricated as per Figure (1) is covered both sides with a thin aluminium foil. The foils should be taut and not touching each other. The two sides of the foil are connected electrically as shown below in Figure 28.4.

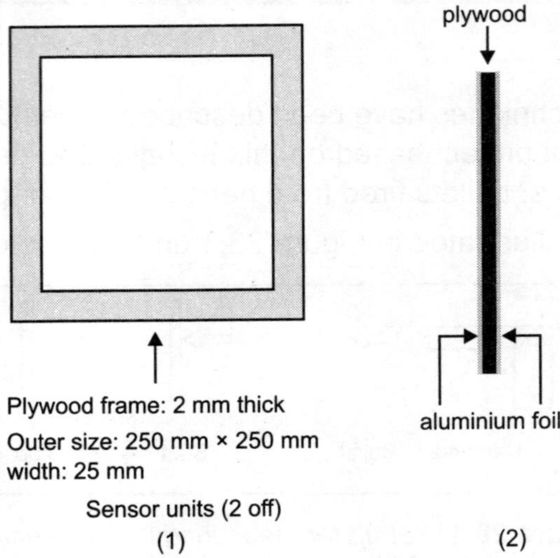

Figure 28.3 *Fabrication of the electro-mechanical sensor*

Operation of the circuit can be understood with reference to Figure 28.2, which shows the trigger arrangement.

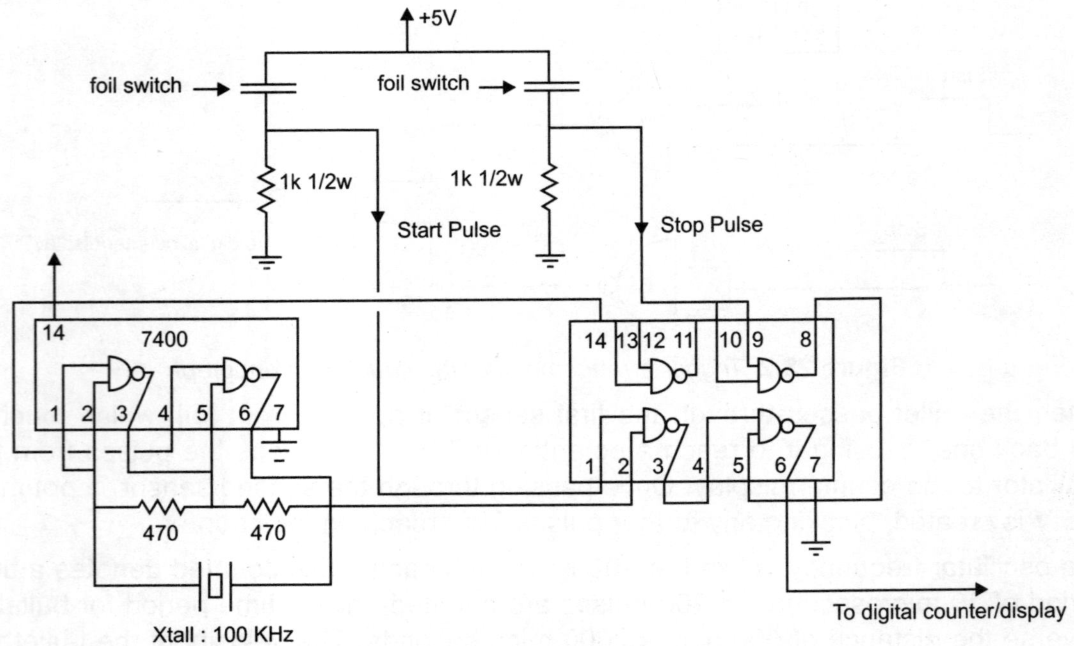

Figure 28.4 *Circuit diagram of the counter chronograph*

From Figure 28.4, we see that the 100 KHz pulse frequency is generated by the crystal controlled oscillator using two NAND gates.

Initially, input 2 of NAND gate A is low; its output is high (Figure 28.2). The pulses from the oscillator are blocked. Similarly, inputs of gate B are also low, but again its output is high. The output from gate C however remains low.

As the bullet hits first sensor, two foils on either sides of the plywood come in contact and input 2 of gate A goes high, causing the pulses from the oscillator to reach the counter. The counting commences at this stage. When the bullet reaches the second sensor, both its inputs 1 and 2 go high, its output goes low, causing the gate C to stop the pulses passing through to the counter. The distance of the first sensor from the gun should be as small as practicable, since we are measuring the muzzle velocity of the bullets, which start slowing down once they leave the guns. The distance between the two sensors is 500 mm, which does not appreciably decrease the bullet velocity while it traverses the distance between the two sensors.

The metal foils used in the sensors should be thin enough so that there is no appreciable slowing down of the bullet as it rips through the foils. The digital counter display is commercially available. The circuit can be assembled on a perfboard. A suitable power supply for this application is given below in Figure 28.5.

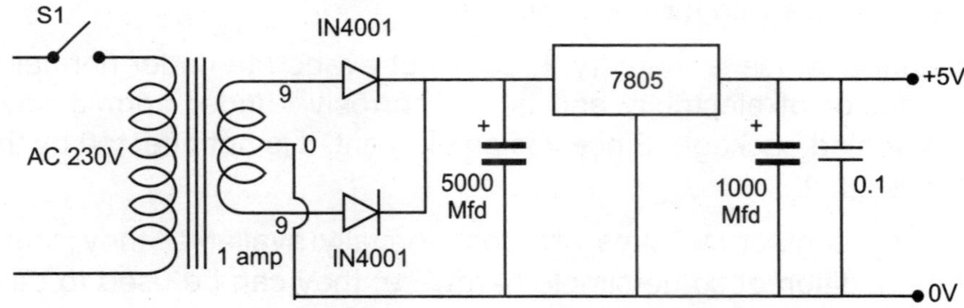

Figure 28.5 *Power supply for counter-chronograph*

29 Tilt Sensing Instrument

There are occasions when a working platform has to be horizontal with the earth's surface. When any liquid is poured over a flat and horizontal surface, it spreads out evenly. If there is a slightest inclination, then the liquid will accumulate in the region which is lower than the others. This can easily be observed by tilting the plate in different directions.

In this project, a small instrument is described which will determine whether the surface is flat, horizontal with the earth's surface, and if not, in which direction it is inclined at. The tilt sensor presented here is designed to be used with its associated electronic circuit and power supply, and will indicate the direction as well as the amount of tilt.

Firstly, we need a liquid medium whose flow will be used to detect any tilt of a surface. The liquid should not be corrosive and should be electrically conductive. Acids and toxic chemicals are ruled out and water is not suitable for two reasons. Firstly, when electricity passed though water, it electrolyses and splits up into hydrogen and oxygen. Secondly, it evaporates with the passage of time.

An ideal substance to use is mercury. It does not evaporate under normal conditions, is a good conductor of electricity and is non-corrosive under normal environmental conditions in a sealed package. Since it is an element, it is not affected by the passage of electricity through it.

Although mercury contact switches are commercially available, they come in single poles. With the addition of some simple hardware, they can be used to detect tilt in a particular direction. To detect tilts in four directions, the sensor described under this project is very useful.

The tilt sensor described below can easily be fabricated with commonly available hardware. See this Figure 29.1 in coloured illustrations on pages 177–184.

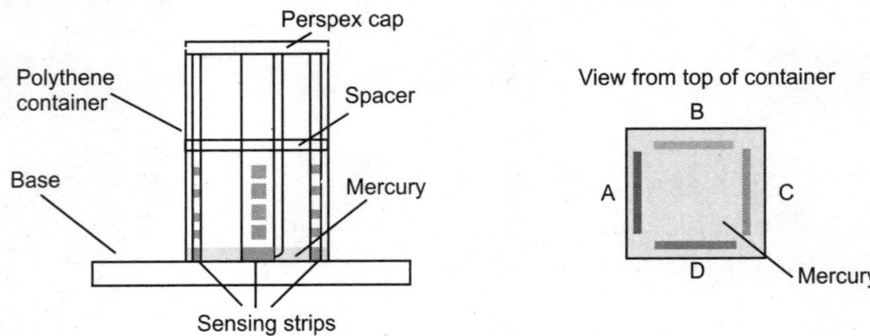

Figure 29.1 *Tilt sensor assembly*

The body of the sensor is fabricated from a square pipe of Perspex or polystyrene with dimensions of 4 cm × 4 cm × 4 cm. One of the ends is sealed off with a similar material. This will be the container of mercury. For sealing/jointing Perspex, chloroform is an excellent solvent.

Four strips of printed circuit boards are prepared as shown in Figure 29.2 in coloured illustrations on pages 177–184. These will serve as electrical contacts with the mercury. The pattern to be etched on copper clad laminates is also given in Figure 29.2. Four such strips will be required for the sensor. The method of etching is described in Part 1 of this book.

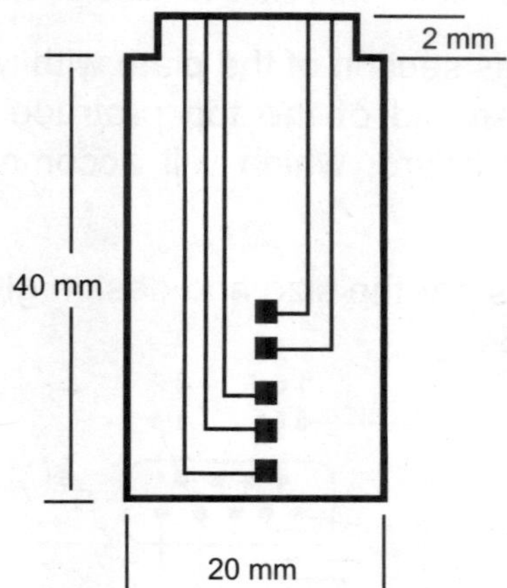

Figure 29.2 *Etching pattern of PCB as sensor strips*

A centre plate called spacer is also made of Perspex, as shown in Figure 29.3 below. This will hold sensor strips in position after insertion of the sensor assembly in Perspex container. The strips pass half way through this centre plate.

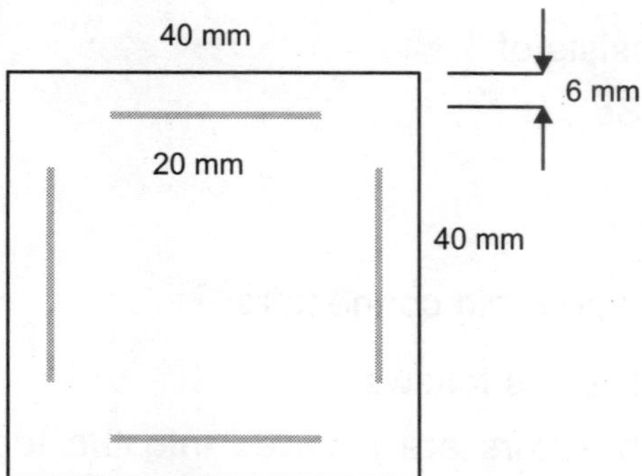

Figure 29.3 *Centre plate also acts as a spacer to separate strips from the container walls*

A top plate, of Perspex is made as per the details in Figure 29.4. This plate will fit over the Perspex cylinder. It will also accommodate connectors. The sensor strips will fit into the cavities shown and protrude slightly to make contact with the top PC board.

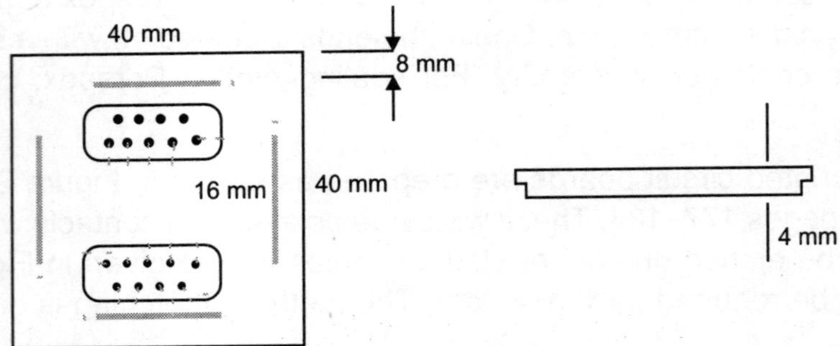

Figure 29.4 *Top plate shown with slots for sensor strips and connector pins*

On the right is shown the cross section of the plate with notches to fit over the Perspex cylinder. The sensor strips should of the top protrude from this top plate equal to the thickness of the top PC board, which will accommodate the two D-type 9 pin connectors.

Fabricate the top PC board as per the size and design given in Figure 29.5 in coloured illustrations on pages 177–184.

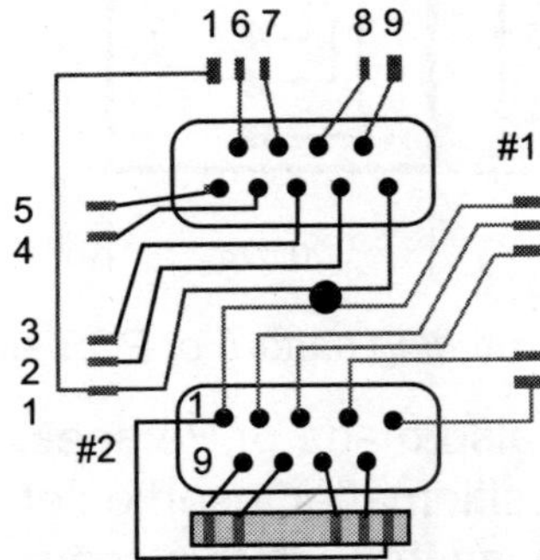

Figure 29.5 *The top PC board: size 40 mm × 40 mm. Etching pattern show above*

The sensor assembly consists of:

(a) Four sensor-strip PCBs

(b) Centre plate

(c) Top plate

(d) Top PCB with two D-type 9-pin connectors

The sensor unit is assembled as follows:

(a) The D-type 9-pin connectors are inserted into the top PCB whose copper side faces downwards, in the allotted position. Solder pins to the PCB.

(b) Insert four sensor strips through the four slots in this PCB and make the connections to the copper lines on the strips and the top PCB through solder. See Figure 29.6 below.

(c) Insert the four PC strip—connector plate assembly through the top Perspex cover plate.

(d) Insert the centre plate from the free end of the PC strips till the plate is half-way up.

(e) Mercury is now added to the square Perspex container till the height of the mercury (on level surface) reaches the top of the lowest PCB line on the sensor strips.

(f) The entire assembly is fixed on to the Perspex container. The top cover should be firmly lodged into the square container. See this Figure 29.6 in coloured illustrations on pages 177–184.

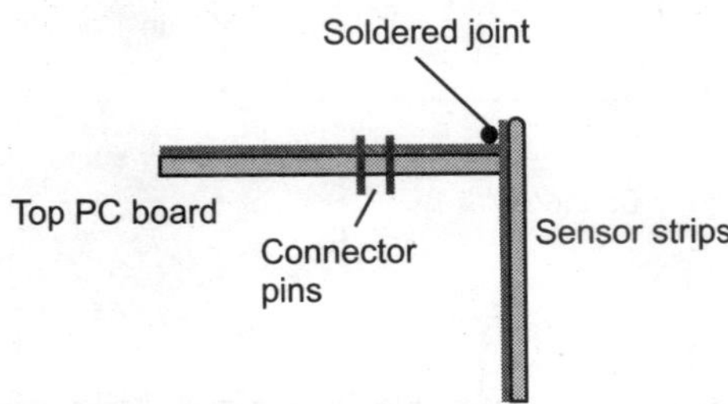

Figure 29.6 *Connecting the sensor strips to the top PCB*

The tilt sensor is now ready for testing.

Use an ohm meter in the low range. Tilt the sensor in one direction so that the mercury now touches PC strip in this direction. With one of the test leads touching pin # 1 of the D connector, a low ohmic resistance should be seen with each of the four levels of the PC strip under test (wherever the mercury is seen to cover the contact). The test should be carried out gradually tilting the sensor more and more till the mercury comes in contact with all four connections on the strip. Repeat this test for other three strips too.

The electronic circuit which will display the information is very simple. It consists of a DC power supply with an output voltage of 3 V at 100mA. The direction and magnitude of the tilt are displayed on a set of 4 × 4 LEDs of four different colours. The instrument is connected to the sensor through a 17-wire cable. The cable from the instrument terminates in two D-type male connectors of 9 pin each.

The complete illustrations of the electronics are given in Figure 29.7 and 29.8.

Figure 29.7 in coloured illustrations on pages 177–184, shows how the sensor strips are connected to the 16 LEDs, which are mounted in the display unit. The connections to the power supply are also shown here. It will be observed that the cathodes of all the LEDs are connected to the negative side of the power supply. The positive side of the power supply, + 2 V, is connected to the lowest contacts of all sensor strips. These contacts will always be electrically connected to the mercury in the sensor unit.

The mercury will connect the positive line to other contacts, depending on which direction and to what extent the tilt is. On a surface parallel with the earth, the mercury will stay confined to the lowest contacts and none of the LEDs will light up.

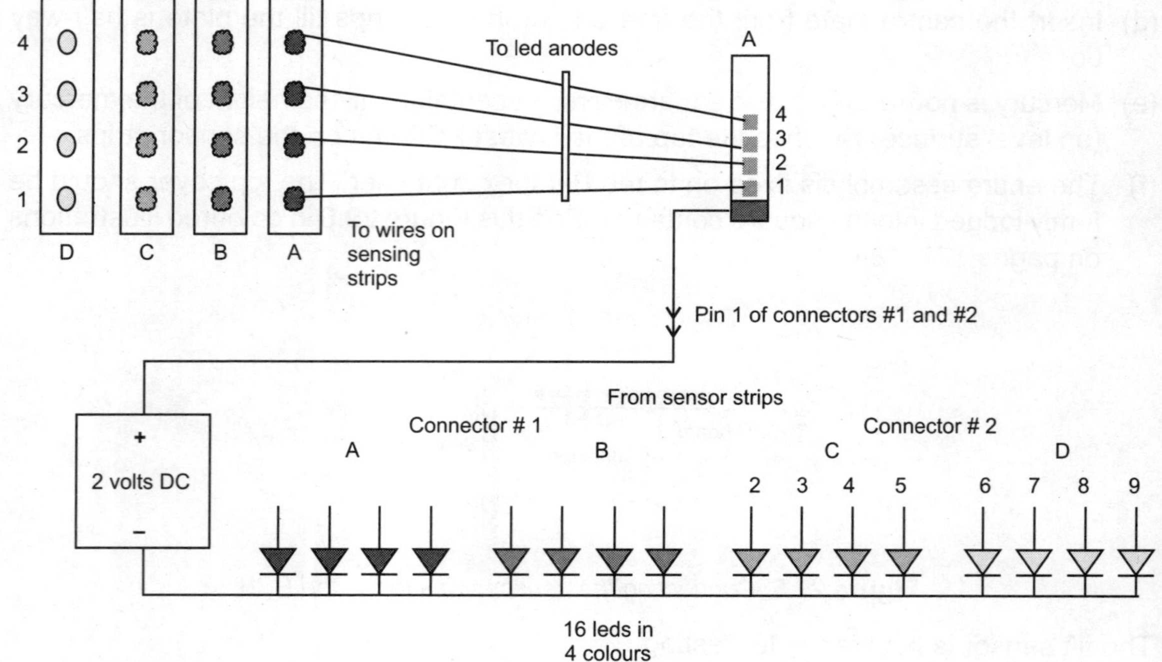

Figure 29.7 *Wiring diagram for the LEDs located on the front panel of the display unit*

The power supply for the tilt sensor is a simple one. A transformer with 6 V output at 150 mA is used with a single diode rectifier, a filter capacitor and a Zener diode. The two resistors in series with the load limit the output voltage to a safe 2-volt level, so that LEDs can be directly connected to the power line (through mercury, of course). See Figure 29.8 below.

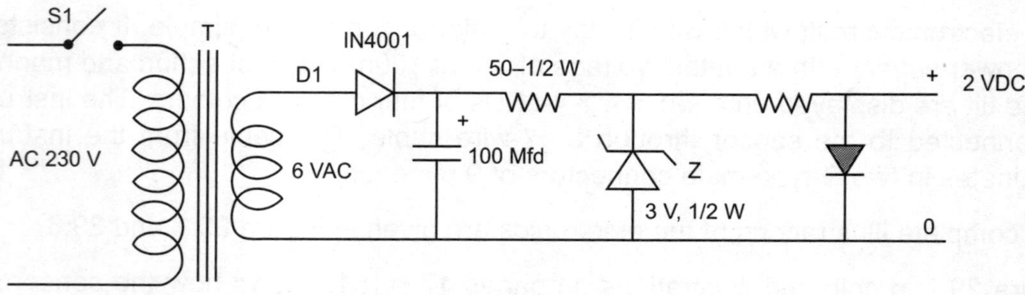

Figure 29.8 *Power supply for the Tilt Sensor display unit*

The power supply and LED display are housed in a small cabinet. The power switch and power-on indicator LED are also located here. The cable from LEDs and power unit to the sensor unit has two 9 pin D-type male connectors, which fit on to the top of the sensor module. This cable has seventeen wires. The other end of the cable is soldered directly to the LED anodes in the cabinet.

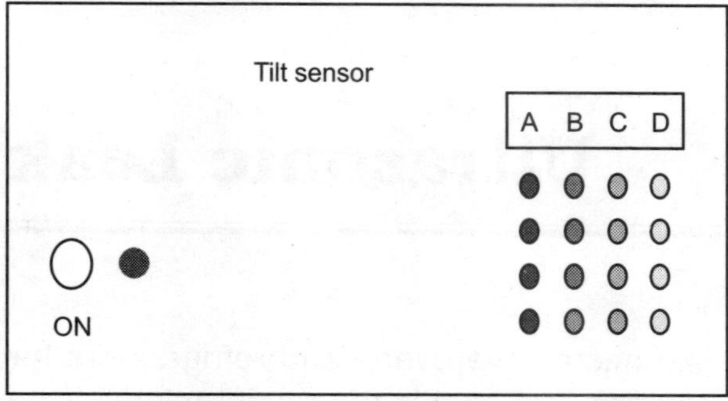

Figure 29.9 Shows the front panel of the complete instrument.

In Use

When the tilt on any surface is to be determined, the sensor unit should be placed on it. To keep the sensor from slipping in case of a tilt, a double sided sticker should be placed below the sensor unit. See this Figure 29.9 in coloured illustrations on pages 177–184.

30 Ultrasonic Leak Detector

Domestic appliances like microwave ovens and refrigerators have an airtight seal. In the case of microwave ovens, the seal is to prevent microwave radiation from leaking out. In the case of refrigerators, the seal is to ensure no leakage of cold air from inside, which decrease efficiency of the appliance. The same situation occurs in cars and other passenger vehicles, which are air conditioned. Defective seals cause leakage of air in all such cases.

An easy to build instrument is described here which uses ultrasonics for leak detection. Audible sound is not suitable for this purpose as wavelengths are large. With the use of ultrasonics though, we have a problem. They are not audible; hence, we have to convert the ultrasonics to audible frequencies using special techniques.

When two sine waves of different frequencies are mixed, two new frequencies are produced. One is the sum of the two frequencies, the other being the difference.

If F1 and F2 are original frequencies, the new frequencies, assuming that Fi > F2, will be:

$$F3 = F1 + F2$$

$$F4 = F1 - F2$$

These are called the heterodyne frequencies and are the basics of superhet radio receivers.

In the current project, we use a small ultrasonic transmitter operating at 40 KHz. To detect this ultrasound, we have a receiver—the output of which is mixed with a local oscillator operating at 39 KHz, resulting in a heterodyne frequency of 1 KHz, which is within the audible range. This activates a pair of headphones. See Figure 30.1 for better understanding.

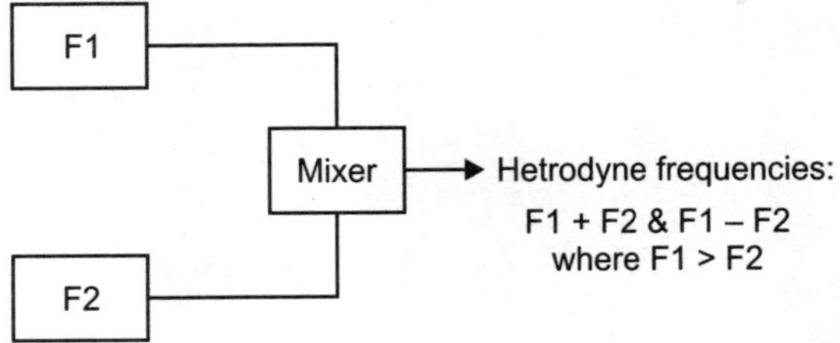

Figure 30.1 *The heterodyne effect—mixing two frequencies*

A block diagram of the complete system is shown in Figure 30.2 below.

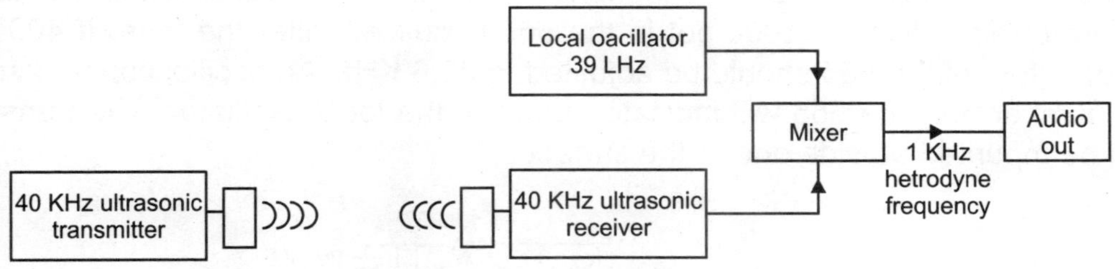

Figure 30.2 *Block diagram of the ultrasonic leak detector system*

Electronic Circuits—Ultrasonic transmitter

The circuit for the ultrasonic transmitter is given in Figure 30.3.

The circuit should be assembled on a small PCB and mounted with the switch in a small enclosure. The transducer should be fixed on one of the outer surfaces.

The circuit for the receiver side of the instrument is shown in Figure 30.4.

The ultrasonic signal is received by the transducer and amplified by two stages of Op Amps, followed by a Schmitt trigger.

The 40 KHz pulses from the Schmitt trigger are mixed with the 39 KHz pulses from the IC 555, which serves as the local oscillator. The resulting 1 KHz signal is fed to another Op Amp, which feeds the headphones marked SP. Power for this circuit is derived from a small 6 volt battery.

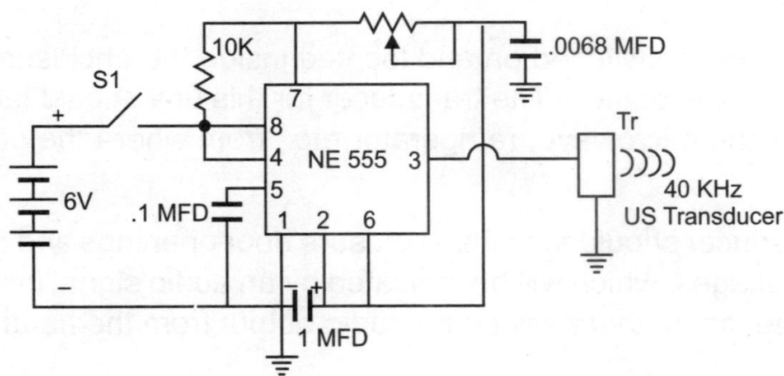

Figure 30.3 *40 KHz portable ultrasonic transmitter*

The 40 KHz pulses from the Schmitt trigger are mixed with the 39 KHz pulses from IC 555, which serves as the local oscillator. The resulting 1 KHz signal is fed to another op amp, which feeds the headphones marked SP. Power for this circuit is derived from a small 6 volt battery. A diode reduces the voltage to 5.4 V, which is connected to different parts of the circuit. Figure 30.4 given below is the circuit for the receiver-amplifier-mixer.

The circuit is assembled on a PCB and house in a small cabinet, together with the batteries. It is advisable to use headphones instead of a speaker. If the 1 KHz tone is not preferred, adjusting 100k pot in the 555 circuit will alter the tone. If 400 Hz is desired, output of the 555 should be adjusted to 39.6 KHz. An oscilloscope connected to Pin number 3 of the 555 will indicate output of the local oscillator. The transducer should be mounted outside one of the surfaces.

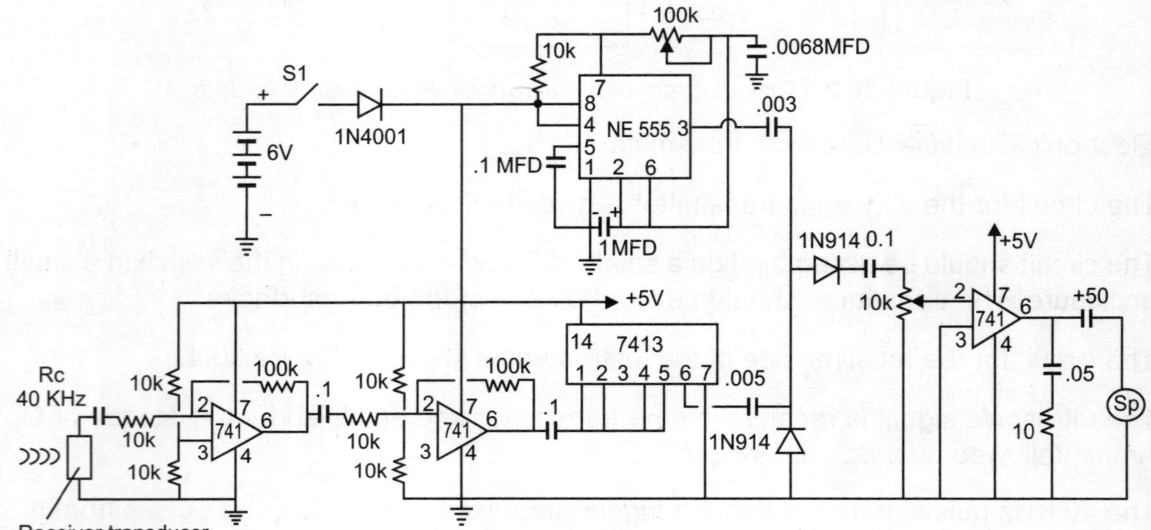

Figure 30.4 *Circuit of the receiver section of the instrument*

The two instruments are meant for portable use, hence they should be physically as small as possible.

In use, the transmitter is switched on and located inside the enclosure from where the possible leaks are to be located. The transducer for this unit should face the outer door of the enclosure – the microwave, refrigerator etc., from where the ultrasound is likely to escape.

The receiver transducer should face the enclosure door openings and moved around for monitoring any leakages, which will be indicated by an audio signal in the earphones. In absence of any leakages, there will be no audio output from the headphones.

PART IV

Major Projects

Differential Delay Analyser 31

A differential delay analyser is an instrument which arranges signals from any source according to their time of occurrence. The source of signals may be a photomultiplier to plot the decay curves of phosphors; a nuclear radiation detector for short-lived radioisotopes; or a circuit generating varying frequencies.

Differential delay analysis forms an important tool in nuclear physics, particularly in neutron physics. Multi-channel instruments are used to sort out the pulses received from a counter according to time intervals in which they occur and recorded in different channels. In such cases, channel widths are of the order of microseconds or even picoseconds.

Channel Widths

The width of the channel is specified in seconds, milliseconds or micro-seconds, as required. This means that each of the channels record the number of events occurring in that particular time duration. To clarify this, consider the following data of audio frequencies from an AF source:

(a) **Total duration:** 1 second, divided into ten parts—each having a duration of 0.1 second.

(b) The channel width here is 0.1 second.

(c) The number of channels is 10.

(d) The number of pulses recorded in the first 0.1 second will be stored in Channel # 1; the number in the duration 0.1 – 0.2 in Channel # 2; and so on.

This instrument is then a ten channel differential delay analyser with channel widths of 0.1 second each.

As an example, consider the decay curve of a phosphor excited by an external radiation, e.g., ultraviolet. One such curve is given below in Figure 31.1.

The source of the signal to be analysed can be any detector or even an audio source. The width of channels defines accuracy—the wider the channel, the worse the resolution. The number of channels multiplied by the channel widths will determine the total time for analysis.

The data from each channel is stored in digital counters. A print-out of the data can be obtained if a printer with an appropriate interface is used.

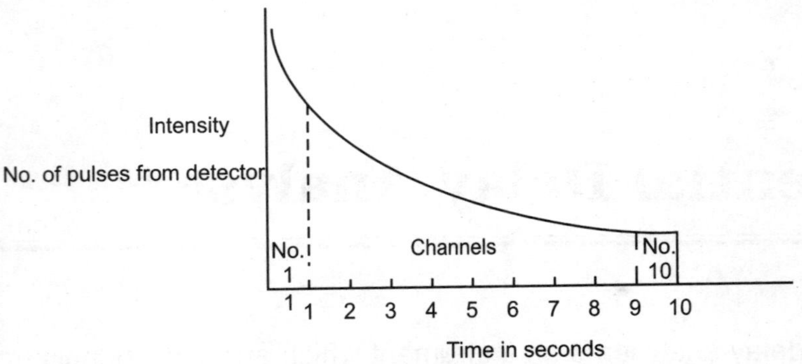

Figure 31.1 *Plot of data indicating a decaying phosphor*

Instruments with wider channel widths can be used for other purposes.

Under this project, we show how to develop a differential delay analyser consisting of four channels. It incorporates an additional facility—the channel widths can be selected from one of the three provisions for this purpose.

A block illustration of the analyser is given in Figure 31.2.

The signal source should be TTL compatible. The pulse oscillator sets width of the channels and pulses are fed to a four stage non-circulating shift register. Each pulse from the oscillator shifts Q output to the next stage. When the fourth stage is reached, Q output from this stage will go to a low level and further action stops after this.

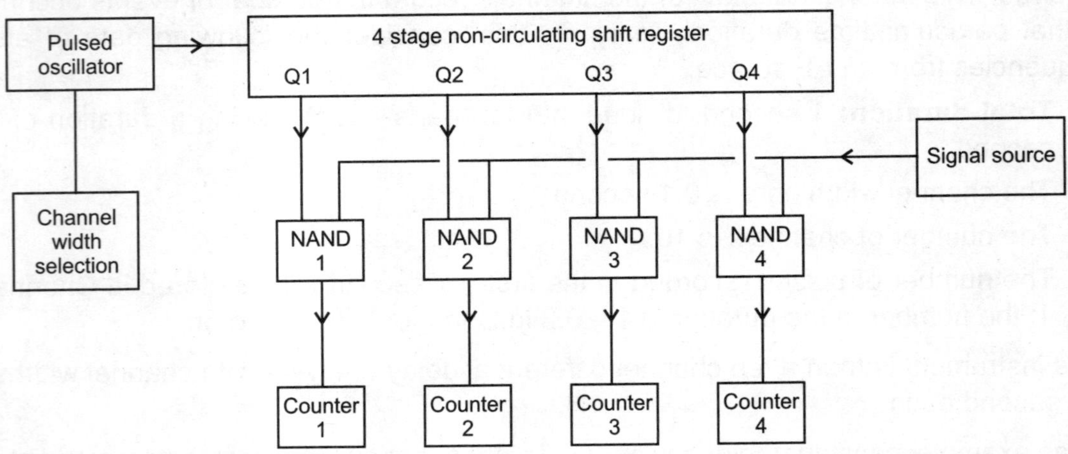

Figure 31.2 *Blok diagram of four channel differential delay analyser*

The four outputs of the shift register are fed to four NAND gates; the other inputs of the NAND gates being tied and connected to the signal input. In this manner, signals from the source pass through each of the NAND gates in sequence, for a duration which is determined by pulse duration of the pulses from the oscillator.

Outputs from each of the NAND gates are fed to a four stage digital counter with a capacity of 1000 counts. If more than a thousand counts are expected, another stage of the counter can be added.

The individual units of the analyser are described hereunder.

Pulsed Oscillator

This is based on Schmitt trigger IC 7413. Only half of the device is used here. With the circuit constants shown in Figure 31.3, we have a choice of three frequencies—1 Hz, 10 Hz and 100 Hz. The pulse duration at 1 Hz would be 1 second; this will be the width of each of the four channels. Since this is a four channel instrument, the total time during which the pulses will be recorded would be 4 seconds.

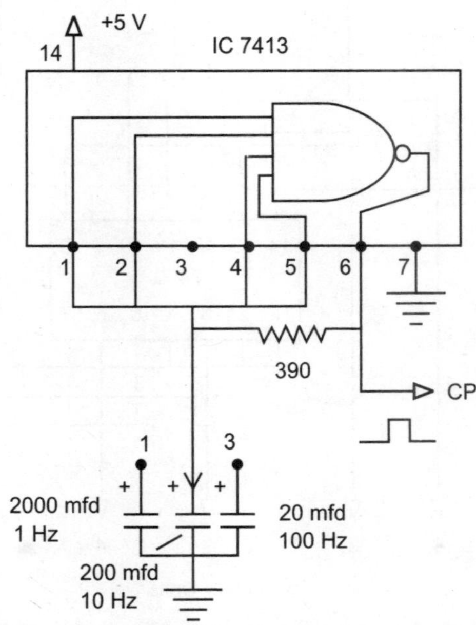

Figure 31.3 *Pulsed oscillator with three pulse durations*

In the switch position 2, the pulse duration would be 0.1 second, this being the channel width with a total of 0.4 seconds of recording available.

In the third position, with a frequency of 100 Hz, a pulse duration – and channel width – would be .001 sec. Tantalum capacitors should be used here and the resistor should have a tolerance Of 1%.

The output pulses are taken from pin 6 of the IC and fed to the four-stage shift register shown in Figure 31.4.

The shift register is based on 7476 ICs, which are J-K flip-flops with preset facility. This enables the first output Q1 to be 0 at reset. Each of the 7476 ICs has two flip flops. We need five such FFs since the fourth one has to be driven to zero level after the fourth pulse, otherwise it will go on recording the pulses if the 1 level stays put. The fifth FF is from the third 7476 IC, the sixth one not being used here.

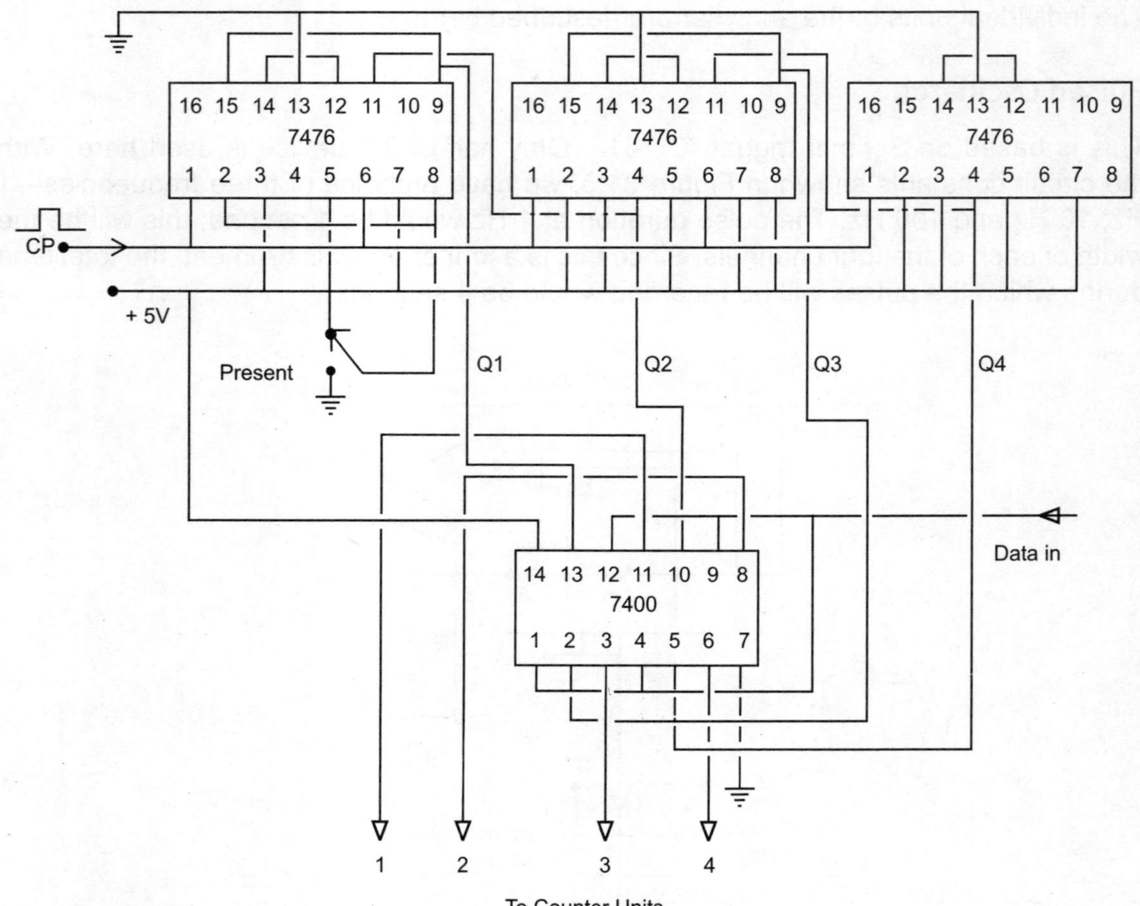

Figure 31.4 *Analyser processing circuit with shift register*.

The truth table for this shift register, which is a circulating type, is given below in Figure 31.5.

As can be observed, if the first Q output of the shift register starts with a '1' level, the remaining outputs will be low. This results in the first NAND gate conducting and pulses from the input, shown in Figure 31.4 as 'Data In', will be passed on to the counter, for a time determined by pulse duration of the oscillator. If the selector switch is in position 1, this will be for one second.

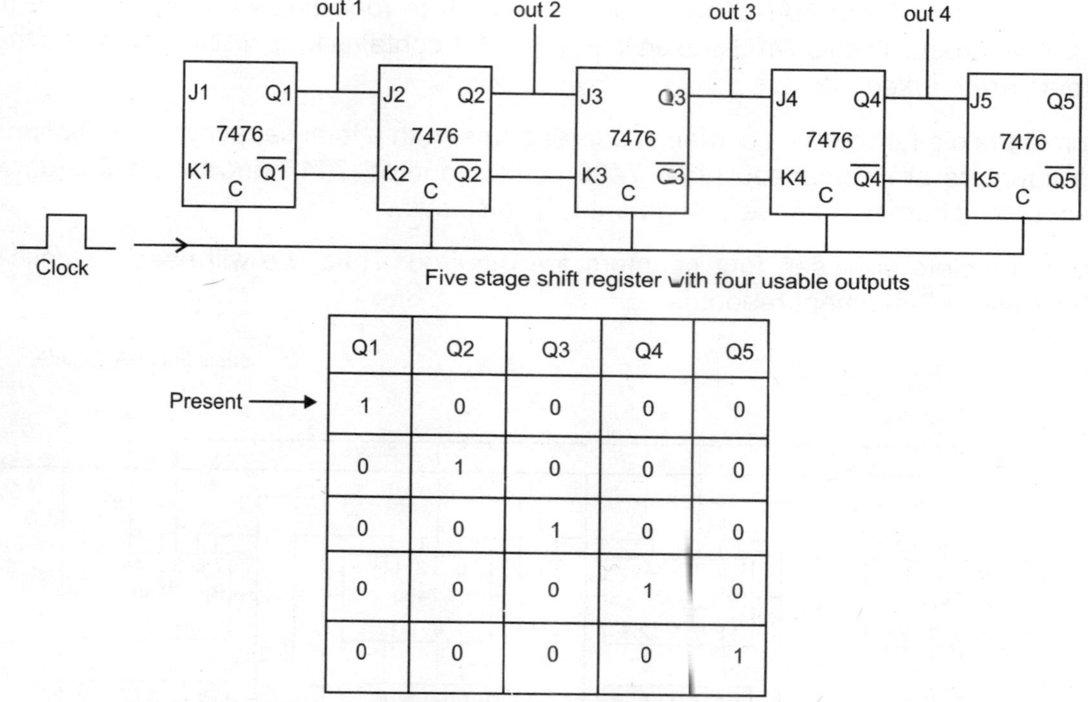

Five stage shift register with four usable outputs

	Q1	Q2	Q3	Q4	Q5
Present →	1	0	0	0	0
	0	1	0	0	0
	0	0	1	0	0
	0	0	0	1	0
	0	0	0	0	1

Truth table showing how the outputs from the shift register shift with each pulse from the oscillatr

Figure 31.5 *Operation of shift registers*

With every pulse from the oscillator, the high Q output will shift to the right as per the Truth Table given in Figure 31.5. Consequently, the pulses will be sent to counters 1, 2, 3 and 4 with each incoming pulse. On the fifth pulse the high Q shifts to FF 5, after which, the action stops since the next FF is not connected for operation. Thus, we will have the time distribution pattern of the signal stored in the counters. The action of the NAND gates in permitting or preventing pulses from the source of the signal to the counters is shown in Figure 31.6 below.

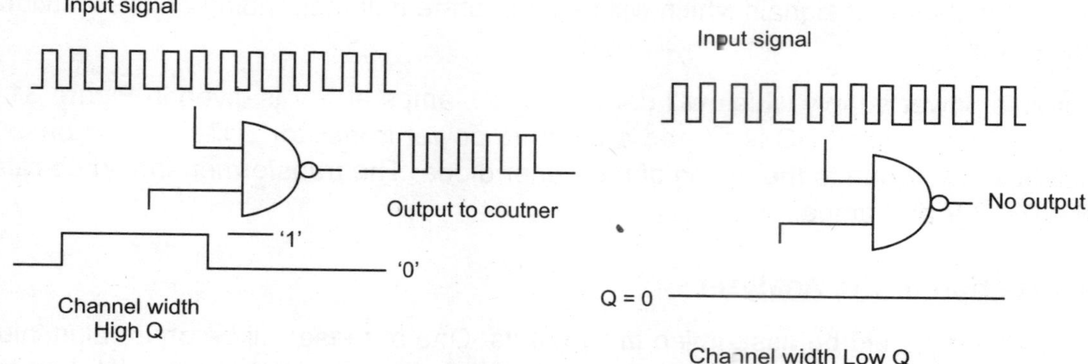

Figure 31.6 *Action of NAND gates in the differential delay analyser*

The outputs from four NAND gates are passed on to four digital counters where the data is recorded. The IC 7400 shown in Figure 31.7 contains four NAND gates and their outputs are marked 1 to 4.

Each of the digital counters contains four readouts, with a total capacity of 9999 counts. The counters are comprised of IC 7490 digital counter, 7447 decimal to 7-segment decoder and common anode 7-segment LED displays.

In the complete analyser, four counters are required. Thus, we will need 12 units of 7490-7447 to 7-segment readouts.

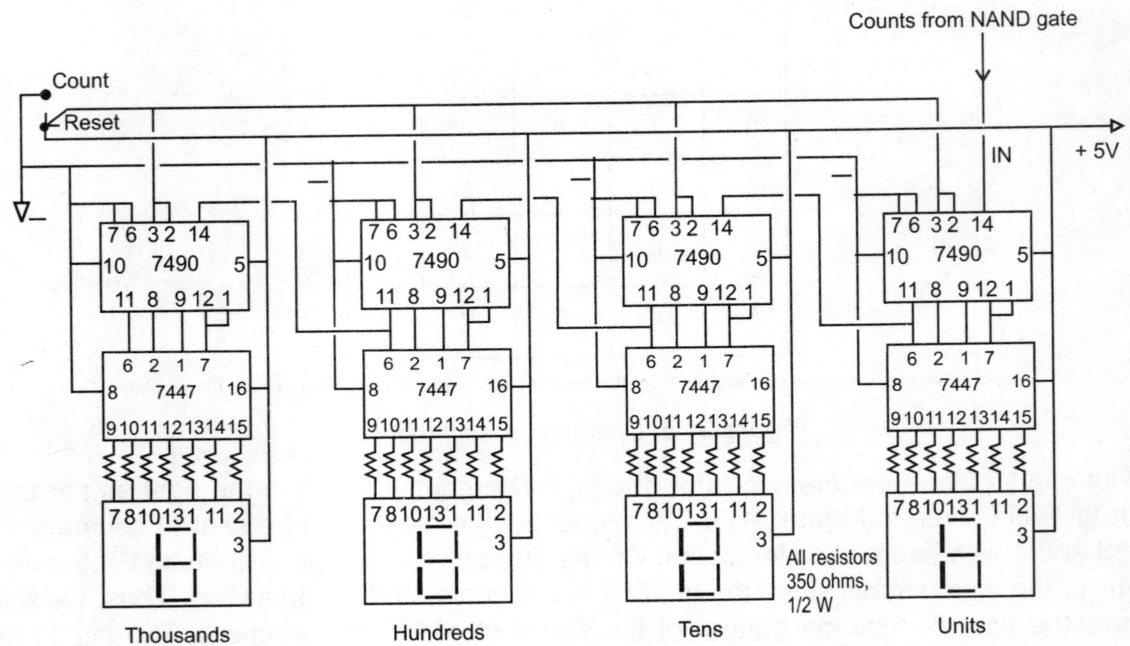

Figure 31.7 *Four stage digital counter for use with the analyser*

A single power supply of 5 V is required for the complete analyser. This does not include the source of signals which will be a separate unit, depending on the individual requirement.

A suitable power supply which will deliver up to 2 amps at 5 V is given in Figure 31.8. The full wave rectified DC is filtered and fed to power transistor 2N3055—the base of which is at 5.6 V due to the action of the Zener diode. The transformer should be rated at 9 – 0 – 9 V at 2 amps.

Construction of the Analyser

The analyser should be assembled in two units. One of these will be on an aluminium chassis and will consist of the power supply and the processing circuit which has five ICs—oscillator, three FF packages and one NAND package. The processing circuit can be assembled on a PC board and fixed on to the chassis, which contains the power supply. The second unit will consist of four counters of four stages each. Each of the counters should be wired separately as an individual module. The four modules can then be assembled with the readouts as shown in the panel layout shown in Figure 31.9.

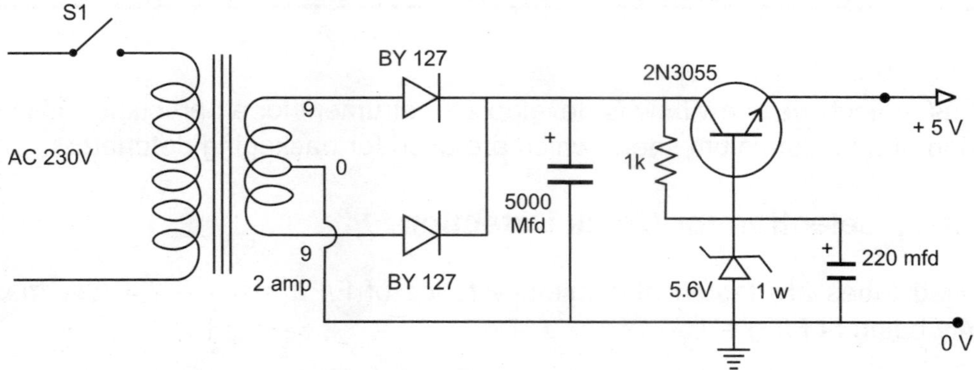

Figure 31.8 *Power supply for the analyser providing 5 V at 2 amps*

The channel width selector switch is a three-position rotary switch. The power and reset switches are toggle switches. The toggle switch for Reset/Count is a DPDT switch, which combines the Reset function for the counter and the Preset function for the shift register.

The input socket is an Amphenol female connector. It should be kept in mind that the input signals are TTL compatible.

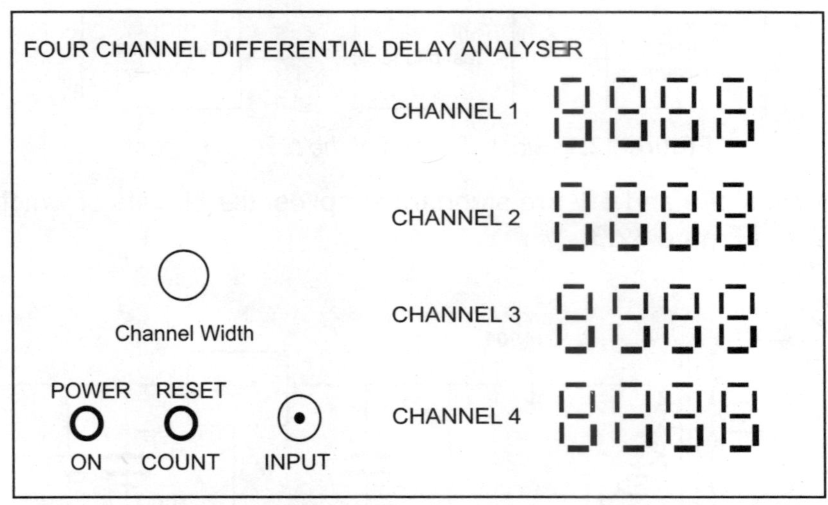

Figure 31.9 *Suggested front panel layout for the analyser assembly*

32 Detecting Defects in Aluminium Shells

Under this project, we show how to develop an instrument for a particular application—the testing of thin aluminium shells which are used for packaging detonators.

Frequency Selection for Crack Detection

Thin-walled tubes are tested at frequency ratios of F/Fg = 0.4 – 2.4, the maximum sensitivity being at F/Fg = 1

The limits of 0.4 – 2.4 mentioned above are for a decrease of maximum test sensitivity to 71 %. In the present case, this sets the limit of test frequencies from approximately 2 –12 KHz, with the maximum sensitivity at about 5 KHz. Figure 32.1 is a block diagram of the instrument.

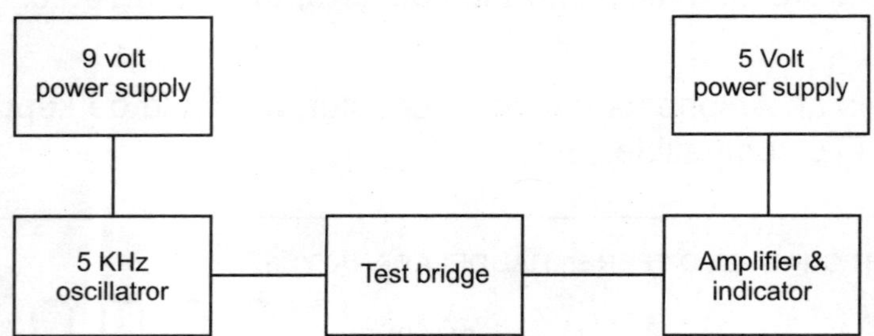

Figure 32.1 *Block diagram of the test instrument*

DC power supplies of 9 and 5 V are standard supplies, the circuits of which are shown in Figure 32.2 and Figure 32.3.

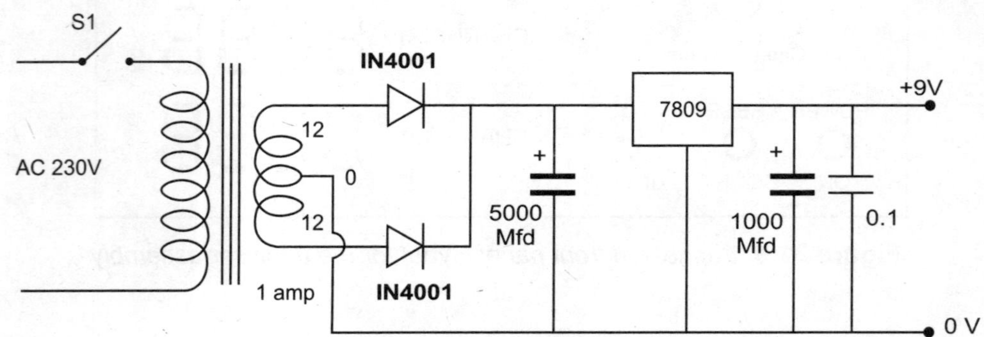

Figure 32.2 *9 V supply for the sine wave oscillator*

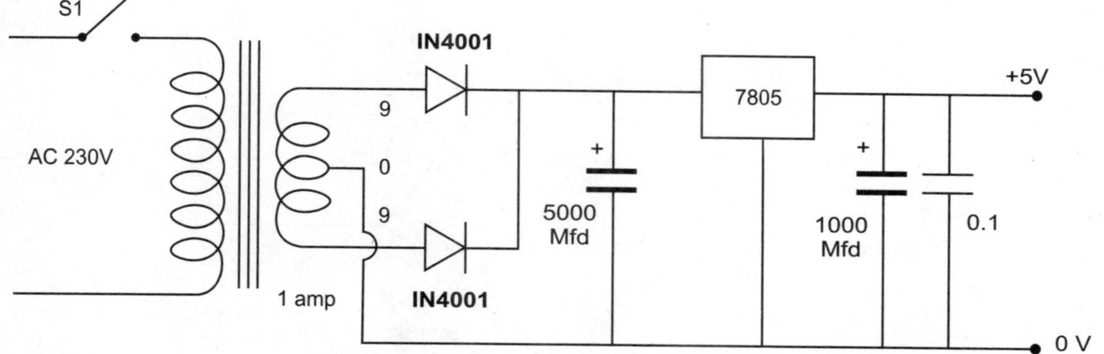

Figure 32.3 *5 V supply for the amplifier section*

Two different power supplies are required because the oscillator and amplifier sections have different zero levels and there cannot be a common earth.

1. The power supplies need centre-tapped transformers to yield full wave rectified outputs to the filter, while two separate ICs—7805 and 7809 provide the 5 V and 9 V stabilised outputs respectively.

The oscillator circuit shown in Figure 32.4 provides a sine wave output signal at 5 KHz for the test bridge. It is based on a Wien bridge configuration.

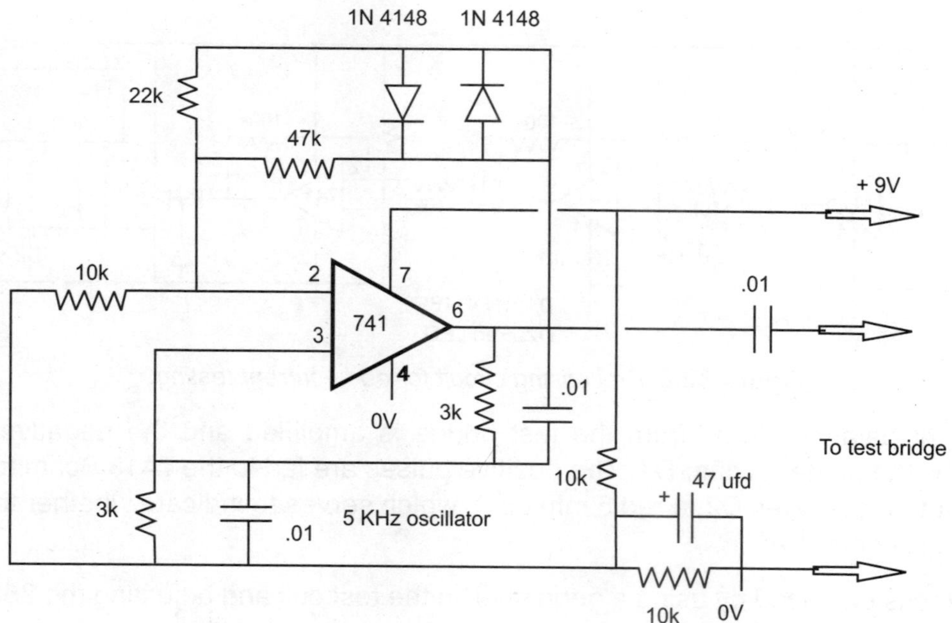

Figure 32.4 *5 KHz sine wave oscillator for test bridge*

The oscillator is based on op-amp IC 741 with the 3k resistors and .01 capacitors being the frequency determining components.

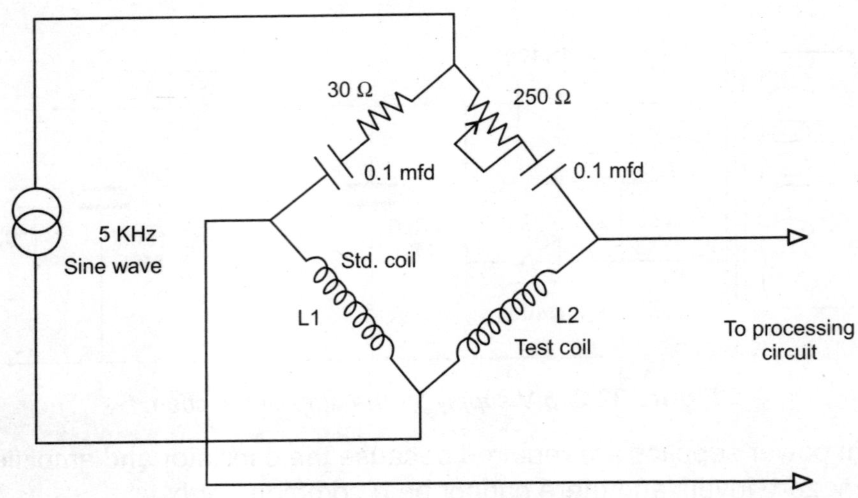

Figure 32.5 *Test bridge circuit for eddy-current tests*

Coils L1, L2, are 3000 turns of # 34 magnet wire wound on formers in which the aluminium shells will fit in snugly.

The output from the test bridge is fed to the processing circuit shown in Figure 32.6. It consists of a three stage amplifier based on 741 op amps.

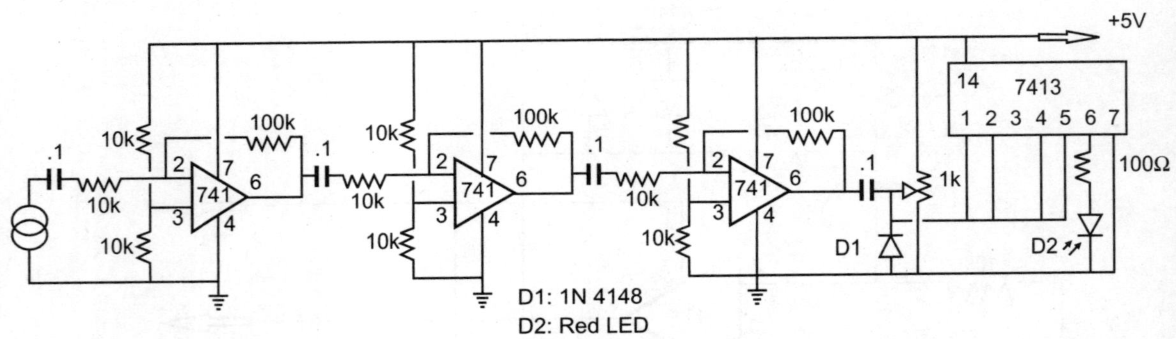

Figure 32.6 *Processing circuit for eddy-current testing*

The out of balance signal from the test bridge is amplified and the negative peaks clipped by the diode rectifier D1. The positive pulses are fed to the 7413 Schmitt trigger whose output activates D2, a red 5 mm LED, which serves to indicate whether the shell under test is defective.

The bridge is balanced by using a good shell in the test coil and adjusting the 250 ohms pot shown in Figure 32.5. The out-of balance signal developed by a defective shell will cause the red LED to light up. The 1k pot shown in Figure 32.6 can be used to adjust the level of acceptable deformity in the shells under test.

Environmental Test Chamber

33

Testing of electronic components, consumables and special materials used in instrumentation requires an environmental test chamber. For routine use, such an instrument has to provide at least two parameters for testing – temperature and humidity. For most consumer based items, a temperature range for testing from 0° C to 100° C is adequate, while a humidity range up to 80% relative humidity is sufficient.

Professional environmental test chambers are very costly. In this project, we describe how such a chamber can be built which will test various non-metallic substrates and insulating material used in a large variety of instruments.

The test chamber can accommodate specimens of material to be tested of sizes 35 mm × 35 mm, the chamber size itself being 75 mm × 75 mm × 75 mm.

The main chamber itself is much larger as can be observed from the dimensional drawings provided herein.

The total size of the main chamber is 300 mm × 300 mm × 225 mm in height, and is made of mild steel. The lid of the chamber at the top is hinged and opens upwards. This chamber accommodates the three power supplies which will feed the various units to control, measure and also perform the insulation tests of the material placed inside the test chamber.

The test chamber, in which the material to be tested will be placed, will have a temperature variation from 0° C to 100° C, while the humidity can be varied up to 80 % of relative humidity.

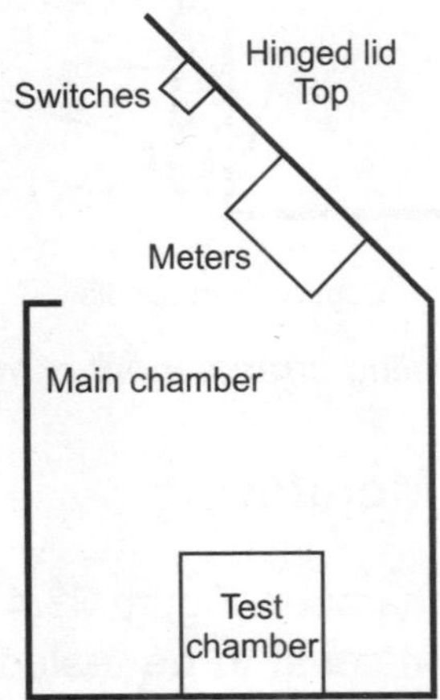

Figure 33.1 *Cross section of main chamber*

Temperature variations in the test chamber are achieved by solid state devices. These are fixed on the side walls of the test chamber – the heating elements on two opposite sides, while the cooling ones on the other two sides. This arrangement ensures an equitable distribution of temperatures as well as varying the temperatures at a faster rate.

Shown in Figure 33.2 – in coloured illustrations on pages 177–184 - is the arrangement of the thermoelectric modules for heating the test chamber.

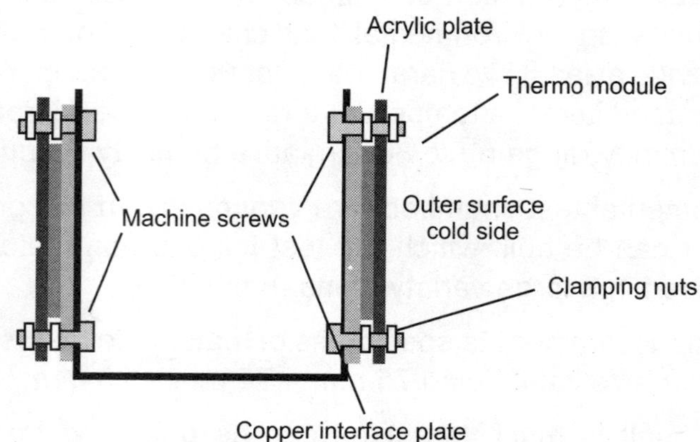

Figure 33.2 *Thermoelectric modules in heating mode*

A temperature sensor and its associated circuit indicates the temperature inside the test chamber – the display being an analogue meter on the top panel of the main chamber. See this Figure 33.3 in coloured illustrations on pages 177–184.

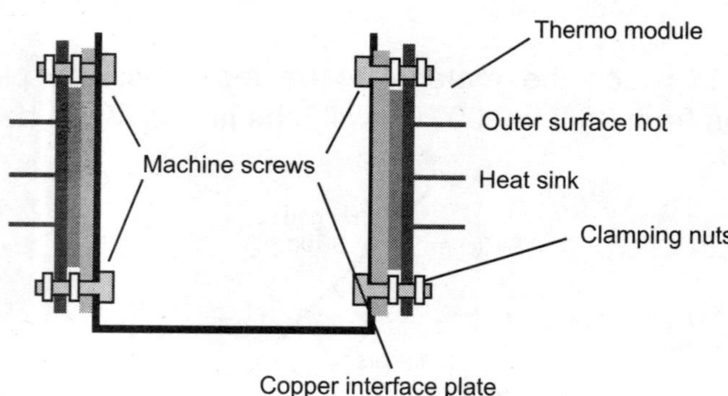

Figure 33.3 *Cooling arrangement of the test chamber*

Temperature Variation and Monitoring

The test chamber inside the main chamber–Figure 33.4 in coloured illustrations on pages 177–184 - will accommodate the material to be tested for their ability to maintain their insulation properties at various temperatures and humidity conditions as specified earlier.

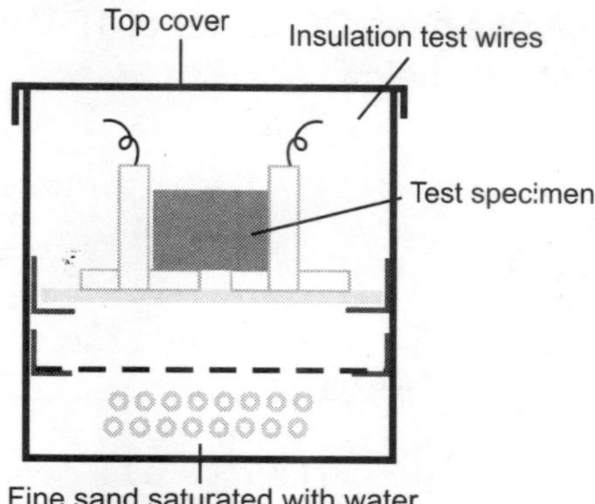

Figure 33.4 *Cross section of test chamber*

For varying the temperature inside the test chamber, a method of heating and cooling has to be incorporated. In this project, we will utilise solid state heating/cooling elements. Two opposite faces of the chamber will heat the unit while the other two sides will cool it. Heating/cooling cycles are selected by a rotary switch located with other instruments on the top panel.

The materials that can be tested for their environmental capabilities include insulation sheets and metals that are painted – to ensure that the paints will withstand the temperatures/humidity conditions so that the metals are protected adequately.

The lower part of the test chamber contains a small quantity of fine sand, saturated with water. When the chamber is heated, the water starts to evaporate, increasing the humidity inside the chamber. In this manner, temperature and humidity tests can be carried out simultaneously. If dry tests are desired, water is not added to the sand. A wire gauze placed above the sand ensures that the sand does not contaminate the test material inside the chamber.

Electronic Instrumentation

The environmental test chamber has three separate power supplies. For the sensor circuits we require a nine V supply for the humidity sensing circuit and a five V supply for the temperature measurement circuit.

A power supply that provides both 9 V and the 5 V is given below.

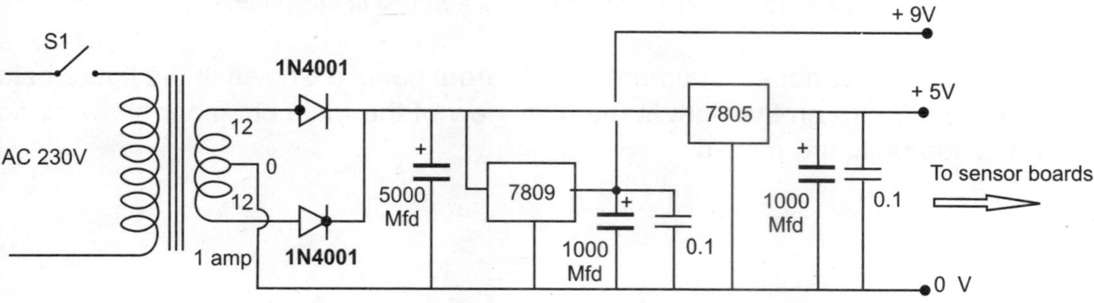

Figure 33.5 *Power supply for the sensor units*

For the thermoelectric modules we need DC power of 12 V at 5 amps. A suitable power supply for this application is shown below:

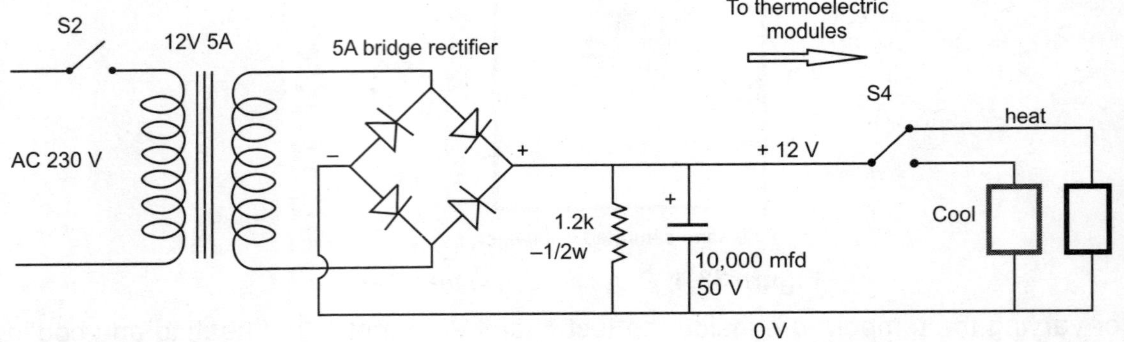

Figure 33.6 *Power supply for thermoelectric modules*

The third and final power supply provides a variable voltage up to 500 V AC for insulation tests of the specimen in the test chamber. The voltage is varied by means of an autotransformer or a simple fan regulator. The test voltage is monitored on an AC voltmeter mounted on the top panel of the main chamber – with the other meters. The AC voltage is fed to the specimen through an AC meter with a range of 0-10 mA, which indicates the leakage current in the specimen at the specified voltage.

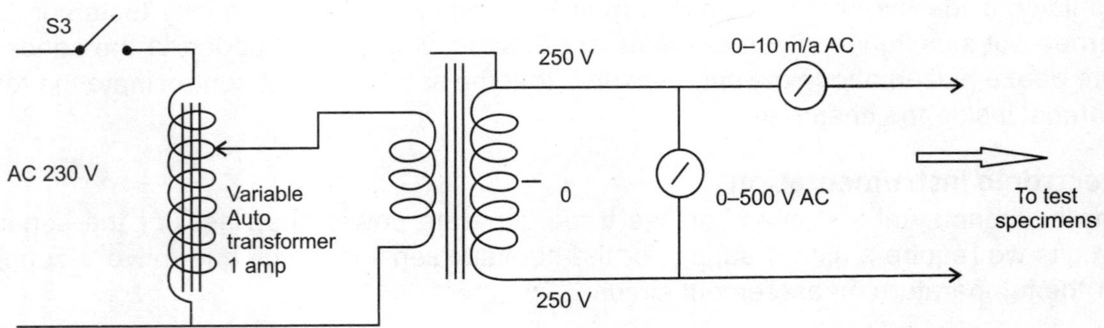

Figure 33.7 *Power supply for leakage test in specimens*

The arrangement of various instruments on the front panel is shown in the figure below on the left. The figure on the right is the plan view of the main chamber showing how the various sub-units are placed.

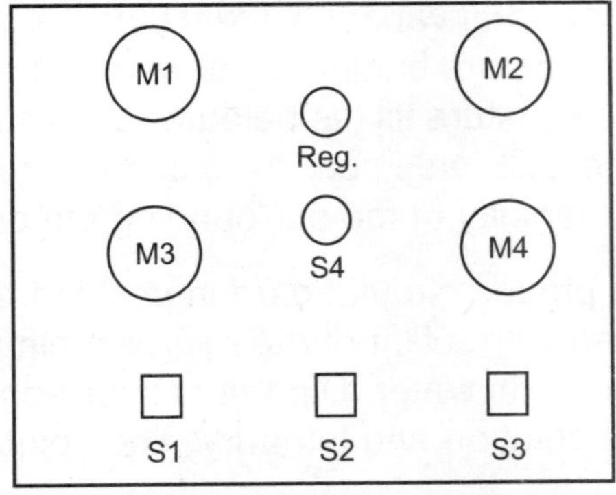

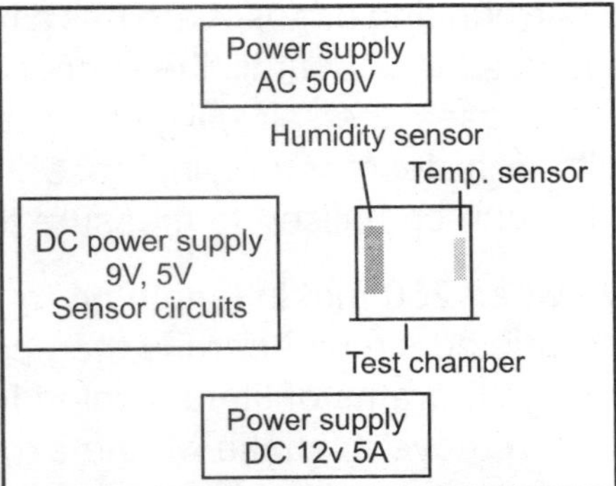

Figure 33.8 *L: Arrangement of instruments on the front panel;*
R: Placement of sub-units in the main chamber.

The figure below shows the circuit of the temperature sensing unit. The sensor is type LM 334, which is designed to deliver 1 u-amp for every degree Celsius.

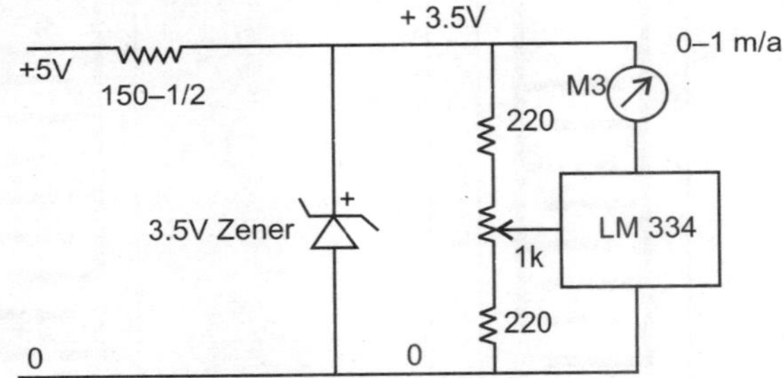

Figure 33.9 *Circuit diagram of temperature sensing unit*

The temperature sensor LM 334 should be fitted inside an aluminium tube with its leads insulated by non-plastic sleeving. This tube is fixed on the underside of the cover that goes over the test chamber. The wire leads from the sensor should be brought out through a hole in the cover plate and wired into the circuit given above.

The temperature sensor should be calibrated using a mercury thermometer with Celsius calibration. In open air, adjust the 1k pot till the current reading in micro-amps equals the Celsius temperature shown on the mercury thermometer. For instance, if the ambient temperature as shown on the mercury thermometer is 34° C, the pot should be adjusted till the meter shows 34 micro-amps. The circuit given in Figure 33.9 should be assembled on a perfboard and fitted onto the power supply chassis.

Humidity Sensor

The humidity sensor has to be fabricated as part of this project activity. It is made of a printed circuit as per the pattern drawing given in

Figure 33.10 below. This will form the base for the sensor. It is basically an electro-chemical sensor which utilises the properties of some hygroscopic substances – salts -, of which lithium chloride is one.

When the surrounding atmosphere has a higher water vapour content than the chemical, the salt absorbs the vapour till the vapour contents are balanced out. When the ambient humidity decreases, the salt will give up some moisture till the balance point is reached. This results in a decrease or increase in the bulk electrical resistance of the salt, a property that can be utilised to measure the humidity of the surrounding atmosphere.

Solder two wires 250 mm in length on to the printed circuit board in position marked 1 and 2. The printed circuit board is then coated with a film of the hygroscopic material. Dissolve about 5 grams of lithium chloride in 10 ml water. Dip the sensor board in the salt solution, remove it, shake off the excess solution and let it dry. The solution must cover the entire copper lines.

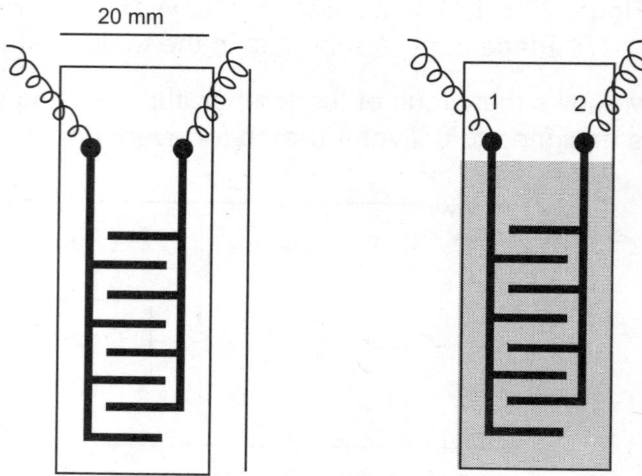

Figure 33.10 *Left, PCB pattern for sensor board. Right: Coated with salt*

An electronic circuit for the sensor operation is given in Figure 33.11 below. This is a bridge circuit with the sensor forming one of the arms.

Since salts suffer from polarisation effects on DC, an AC source should be used with a bridge circuit. In the circuit of Figure33.11, a sine wave oscillator delivering 400 Hz is used. This is generated by a Wien bridge oscillator based on 741 op-amp.

As the environmental humidity changes, so does the resistance across the sensor, the out of balance current being amplified by a 2N3711 transistor.

The sensor has to be calibrated with known values of relative humidity. This is done by using a wet and dry bulb thermometer. Temperature readings are obtained for a dry thermometer and repeated for a wet thermometer. This has to be done at different temperatures. By using standard tables, the relative humidity can be observed. Check the readings with the sensor unit readings, using the pots in the circuit to set the minimum and the maximum limits.

The humidity sensor is then mounted on the underside of the test chamber cover plate, with the leads coming out of a hole and connected to the bridge circuit as per the circuit shown in Figure 33.11.

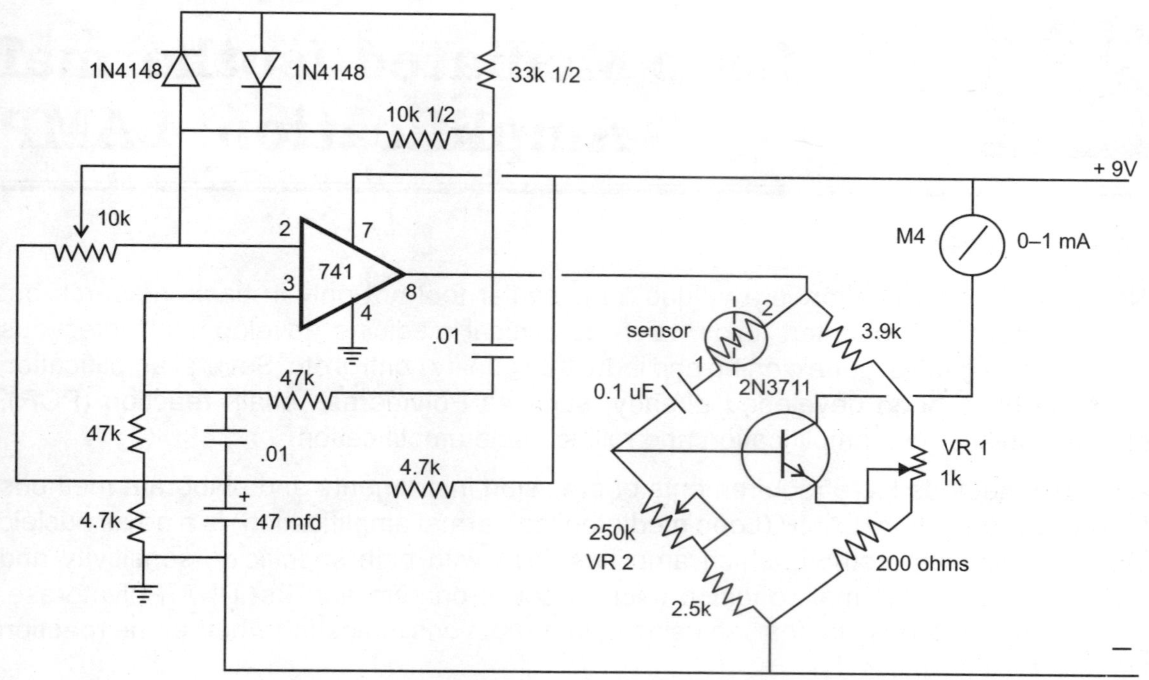

Figure 33.11 *Humidity sensor electronics circuit*

Using the Environmental Test Chamber

The sample to be tested is inserted between the two metallic u-channels inside the test chamber. This provides the connection to the AC voltage source with a capacity to test the samples up to 500 VAC, observing the leakage current through the AC milliammeter. When the meter starts indicating any leakage, the test should be halted and the test voltage noted.

The test should be carried out at different temperatures and varying degrees of humidity; the temperature variations being done by switching on the heating/cooling modules mounted on the test chamber. With the sand saturated with water, the ambient humidity will rise when the test chamber is heated, due to the evaporation of water from the sand reservoir.

If only dry tests are required, the sand should be removed from the test chamber.

The test chamber is inserted into the main chamber – as far away from the power supplies as possible. A small fan installed inside the main chamber will keep the air circulating and prevent overheating the power supplies and the electronic circuits.

34 Loop Mediated Isothermal Amplification: LAMP

Nucleic acid amplification is a valuable molecular tool not only in basic research but also in application-oriented fields, such as clinical medicine development, infectious diseases, diagnosis, gene cloning and industrial quality control etc. Several amplification methods have been developed already, such as Polymerase chain reaction (PCR), strand displacement amplification and rolling circle amplification.

All these methods have requirements of precision instruments and elaborate methods for product detection. LAMP (Loop mediated isothermal amplification) is a novel nucleic acid amplification method, which amplifies DNA with high specificity, sensitivity and rapidity under isothermal condition using a set of primers and Bst DNA Polymerase. The products are easy to analyse using colour reactions indicating whether the reaction is positive or negative.

LAMP relies on auto-cycling strand displacement DNA synthesis which is carried out at 60°–65° C for 2 hrs in the presence of BSt DNA polymerase, dNTP's., specific primers and the target DNA template. The LAMP method employs a DNA polymerase with high strand displacement activity and a set of four specially constructed primers (two inner and two outer primer) that recognise six distinct sequences on the target DNA. The inner primers are called the forward inner primer (FIP) and the backward inner primer (BIP) and each contains two distinct sequences corresponding to the sense and antisense sequences of the target DNA. One of the inner primer initiates LAMP reaction and the other is used for self-priming in later stages.

The mechanism of the LAMP amplification reaction includes three steps: production of starting material, cycling amplification and elongation, and recycling. All stages of the reaction are carried out at a single temperature of ~ 65°C

LAMP is best suited for present / absent type of investigation such as testing for presence of pathogens in a biological sample or the presence or absence of a certain genomic region etc. many such assays for the detection of pathogens have been developed.

The starting material for LAMP is DNA isolated from biological specimens

DNA	=	100 ng
LAMP Primers	=	40 p mol Loop primers and 5 p mol of outer primers
dNTP	=	10 mM
Bst Polymerase buffer(2X)	=	12.5ul
Bst polymerase	=	8 units

Indicator Dye in minimal concentration

The total reaction volume is 25 ul

The reaction mix as assembled above is incubated at a constant temperatures of 65°C for 2 hrs.

The tubes are removed at the end of the reaction and inspected for colour change/ turbidity or alternately the product can be visualized on a gel (Agarose/PAGE).

Development of Instrument for Incubation of PCR Tubes

The PCR tubes have to be inserted into a metallic block which is heated and maintained at a constant temperature for a fixed duration of time.

The tubes will fit inside the cavities in the metallic block, the cavities being of 7 mm diameter with a depth of 20 mm. Dimensions of the metallic block are given in Figure 34.1 in coloured illustrations on pages 177–184.

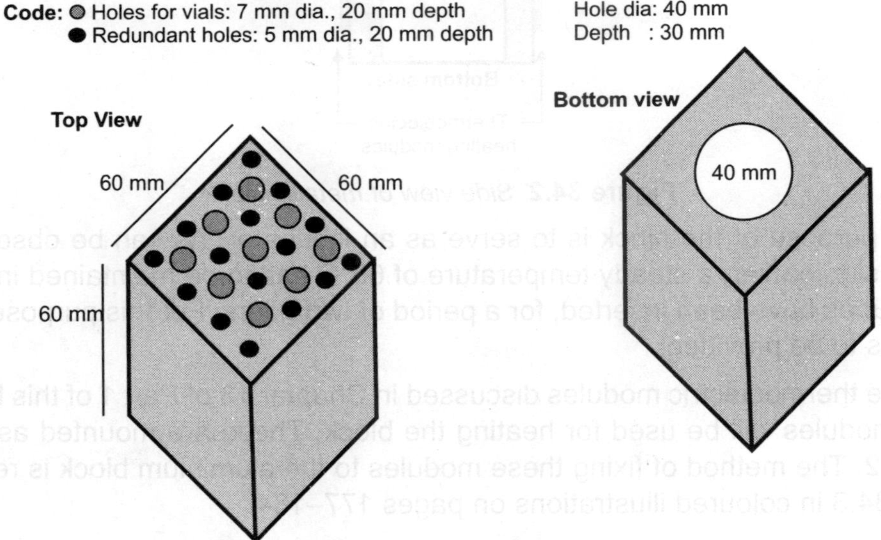

Figure 34.1 *Two views of metallic block for heating vials*

The block is fabricated from solid aluminium square rod of 60 × 60 mm. This results in a cube of dimensions 60 × 60 × 60 mm.

If the block is specially cast from a wooden pattern, then the 40 mm hole in one side should be made in the pattern before casting. This will not only save wastage of material, but will also make the machining easier.

On the opposite surface of the 40 mm cavity, nine 7 mm holes are drilled up to a depth of 20 mm, as per the pattern suggested in the figure above. The holes will accommodate the standard PCR vials.

Another set of sixteen numbers of 5 mm diameter holes are also drilled up to 20 mm in depth. The arrangement of all the twenty-five holes is given in Figure 34.1 above.

This completes the preparation of the main block in which the tubes will fit.

The reason why the additional holes are provided in the block is that as much of unnecessary material as possible should be removed from the block. This will not only save energy in the heating process but will also result in quicker heating and cooling cycles when the same block is used for polymerase chain reaction (PCR) process.

Also shown in Figure 34.2 is in coloured illustrations on pages 177–184, the arrangement of the thermoelectric heating modules, facing each other on the remaining sides of the block.

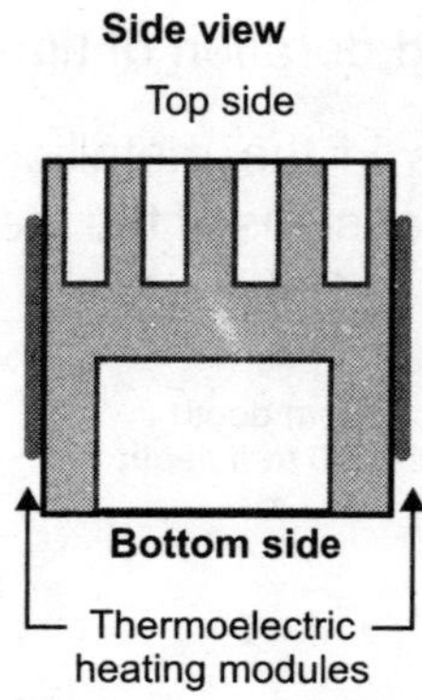

Figure 34.2 *Side view of metallic block*

The main purpose of the block is to serve as an incubator. As can be observed from the discussion earlier, a steady temperature of 65°C has to be maintained in the block after the tubes have been inserted, for a period of two hours. For this purpose a source of heat has to be provided.

We will use thermoelectric modules discussed in Chapter 13 of Part 1 of this book. Two of these modules will be used for heating the block. These are mounted as shown in Figure 34.2. The method of fixing these modules to the aluminium block is reproduced in Figure 34.3 in coloured illustrations on pages 177–184.

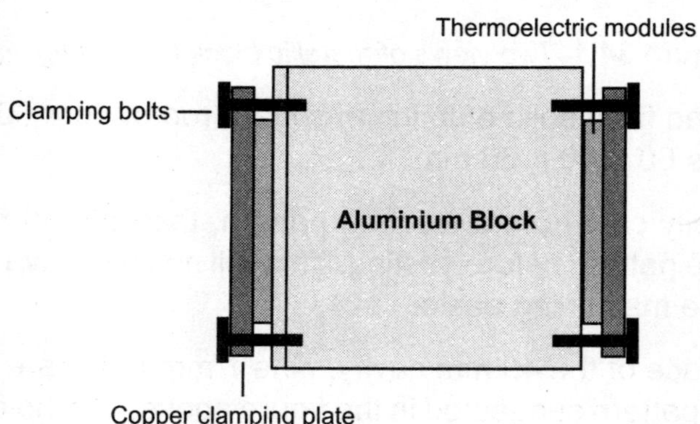

Figure 34.3 *Clamping of heating modules*

For heating the aluminium block, the cold surfaces of the modules (when connected to the power source) should be on the outer side. The modules are clamped tightly between a copper clamping plate and the aluminium block.

The copper plate is held in position by four bolts screwed into threaded holes. In this manner, no damage is done to the heating modules. This is clarified in Figure 34.3.

The heater assemblies should be fixed symmetrically on the side walls of the metallic block for equitable distribution of heat.

There are nine holes of 7 mm diameter on the top surface of the metallic block. Eight of these can be used for inserting the PCR vials, while the ninth – the central one – will accommodate the temperature sensor, which is a silicon diode.

The two connecting wires from the sensor and four from the heating modules will be brought out from the top lid and fixed to a strip connector on the lid itself. This is shown in Figure 34.4.

The aluminium block should be placed inside a container with a lid on the top. Fibre glass insulation should be provided on all the four sides and the bottom part. The top lid should have holes to bring out the wires from the heaters and the sensor.

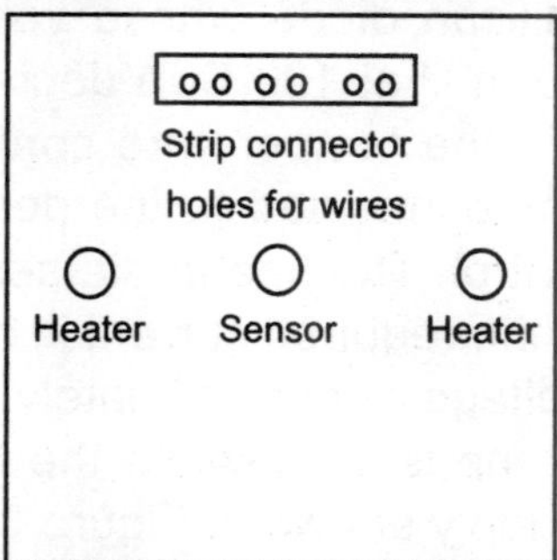

Figure 34.4 *Top lid of the incubator block*

These will be inserted into the strip connector on the lid as shown in Figure34.4 above. The container for the block should be at least 160 mm on each side and 130 mm in height. This will accommodate 50 mm thick fibre glass insulation on all the four sides and the bottom part. Between the top of the metallic block and the lid of the container, a square piece of bison panel should be placed, to conserve the heat from the metallic block.

Electronics Instrumentation of the Incubation Chamber

As mentioned above, the electronics circuits will be required to heat the thermoelectric modules, and to monitor and maintain its temperature at a steady 65°C for two hours.

The Heater Elements

These are semiconductors whose specs are given below:

Specifications of two types of thermoelectric modules are provided below:

ambient = 30C					
Item	Description	Max Delta	Max	Max	Max
Part No		T	Voltage	Current	Wattage
TEC1-12705	Single stage TEC	68	15.4	5.5	49
TEC1-12708	Single stage TEC	68	15.4	8.5	76

The maximum voltage that can be applied to these two types is 15.5 V and the maximum current drawn at these voltages is 5.5 amps for the 12705 and 8.5 amps for the 12708.

In Figure 34.6 is shown a circuit diagram of a simple power supply that will deliver 15 V at 10 amps, which is sufficient for the type TEC1 – 12705, which will be used in this project. With the autotransformer the voltage can be set to 15 V and monitored on the voltmeter.

The temperature sensor is a silicon diode whose voltage drop across its ends will vary from 600m V at 0˚C to 400 m V at 100˚C, a decrease of 200 m V over 100˚C. It is this property that is utilised in the temperature controller whose circuit is given in Figure 34.5. The pot of 5k ohms connected to the positive input of the op amp is to set the required temperature control. The heater elements are powered through power MOSFETS which can handle all the required current in one device. For the temperature sensing network, a reference voltage of approximately 7 V is required. Apart from this the voltage required for the opamp is 12 V while the heater modules need 15 V. All these are provided in a single supply shown in Figure 34.6.

It should be kept in mind that the wires from sensor diode and the heaters will be subject to a continuous temperature of 65˚C over a period of two hours. Hence it should be ensured that the insulation of the wires should be able to withstand these temperatures.

The sensor diode should be placed vertically inside the central hole in the metallic block with the two leads coming out from the top. Ensure that the leads are properly insulated so that they do not come into contact with the metal body of the block.

Temperature Setting

Insert a mercury thermometer inside one of the holes in the metallic block. Connect a 15 amp meter between the positive end of the modules and the 15 V supply line.

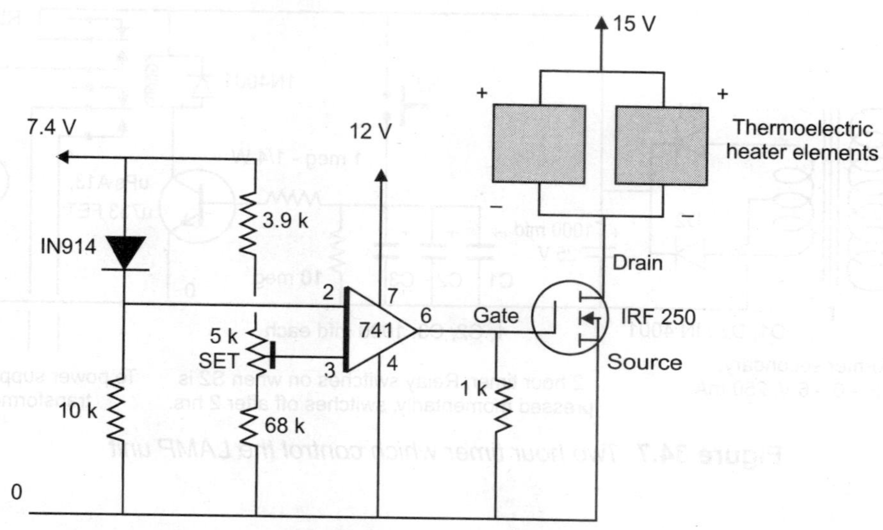

Figure 34.5 *Temperature controller circuit for the incubator*

Switch on the power supply and adjust the temperature adjustment 5k pot so that the meter shows a reading of 15 amps. When the thermometer reads 65°C, adjust the pot so that the current just switches off. With a few trials, the setting can be fixed accurately.

The power supply diagram is given in Figure 34.6 below:

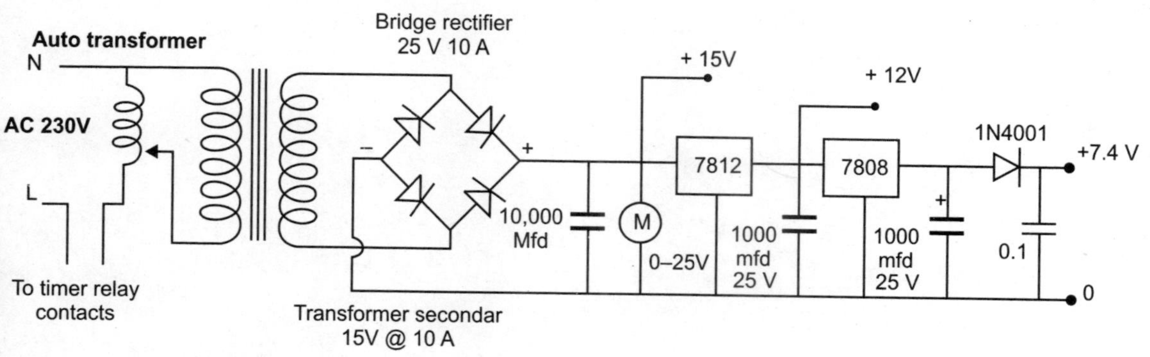

Figure 34.6 *Composite power supply for the incubator unit*

The mains input voltage to the auto transformer is switched through a timer. A timer circuit is shown in Figure 34.7.

When the timer is switched on, the relay contacts close and the composite power supply is switched on. A red LED indicator across the relay contacts also switches on. The timer operates for two hours after which it switches off the AC supply to the composite power supply unit.

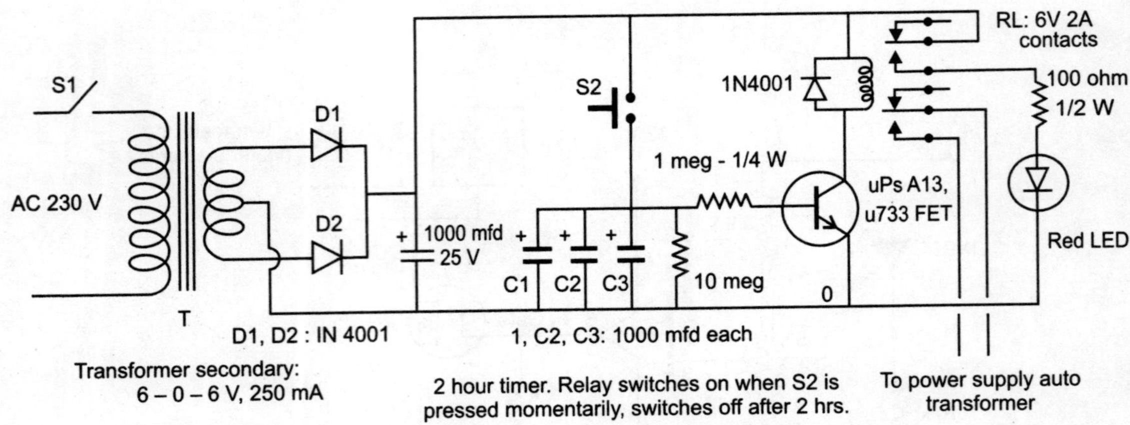

Figure 34.7 *Two hour timer which control the LAMP unit*

Ultrasonic Testing of Solids

35

A variety of electronic techniques are available for measurement of sonic velocities in solids. These have applications in determining their elastic constants. Under this project we take up the development of an instrument to measure the sonic velocities of small samples of solids – metals and non-metals using digital techniques.

The block diagram shown in Figure 35.1 illustrates the principle of the instrument.

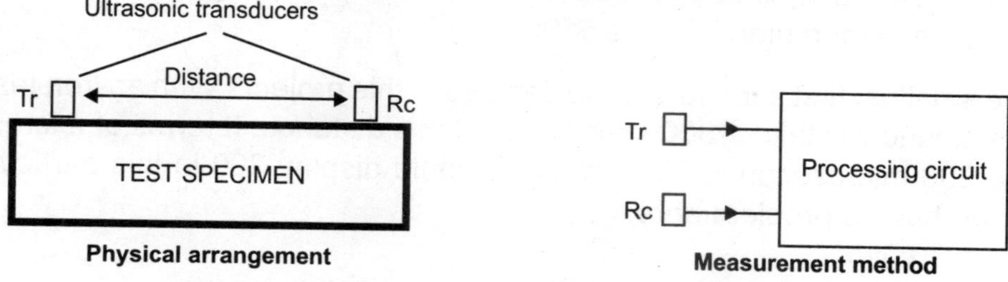

Figure 35.1 *Operating principle of the instrument*

The diagram on the left is the mechanical arrangement of two ultrasonic transducers which are in physical contact with the material under test, at a distance D between them. The processing circuit measures the time taken for the ultrasonic pulse to travel the distance 'D', from which the velocity can be determined. There are two alternate approaches to develop the processor, both of which are described below.

The first method is to use an electronic timer to measure the time 'T' taken for the pulse to travel the distance 'D', then calculate the velocity V = T/D.

The second approach is to use a microcontroller which displays the time directly on a LCD. This method utilises a software programme and one such programme in 'C' is also given later.

We will first describe the method using a separate timer. The diagram in Figure35.2 outlines this.

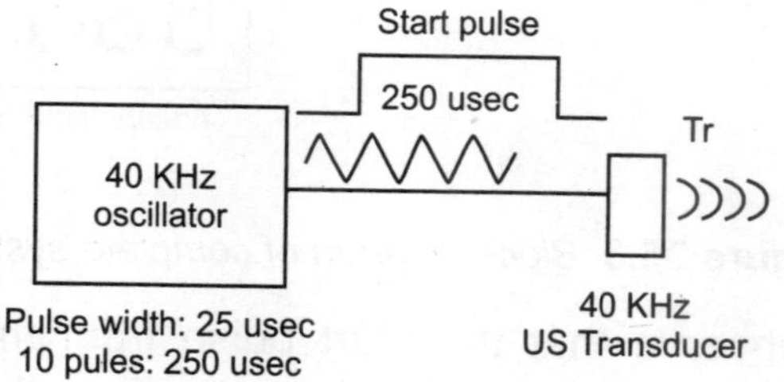

Figure 35.2 *Ultrasonic transmitter: 40 KHz frequency*

The first part consists of an ultrasonic transmitter operating at 40 KHz. This gives a pulse width 25 microseconds. We deliver ten such pulses from the transmitter end, covering a period of 250 microseconds. A start pulse covering this period injects ten pulses into the sample under test.

On the receiver side, a similar transducer picks up the ten pulses for processing the time of transition of the pulses through the specimen under test and displays this time on a digital display. The arrangement is shown in Figure 35.3.

The duration for the ultrasonic pulses to travel the distance between the two transducers placed on the material under test is divided into intervals of one microsecond each; provided by the 1 MHz pulse generator. In this manner, the count displayed on the digital readout is just the duration in microseconds.

The velocity of sound in air at 0˚C is 331.3 m/sec. For iron and glass it is much greater, 5000 m/sec, while for aluminium it is 5100.

What is the smallest test sample that we can use in this project? With an iron test piece 1 meter long, sound will take 1/5000 sec to travel this distance. It terms of microseconds, this will be 200 microseconds. The counter should display 200 in this particular case. However, we have a problem here.

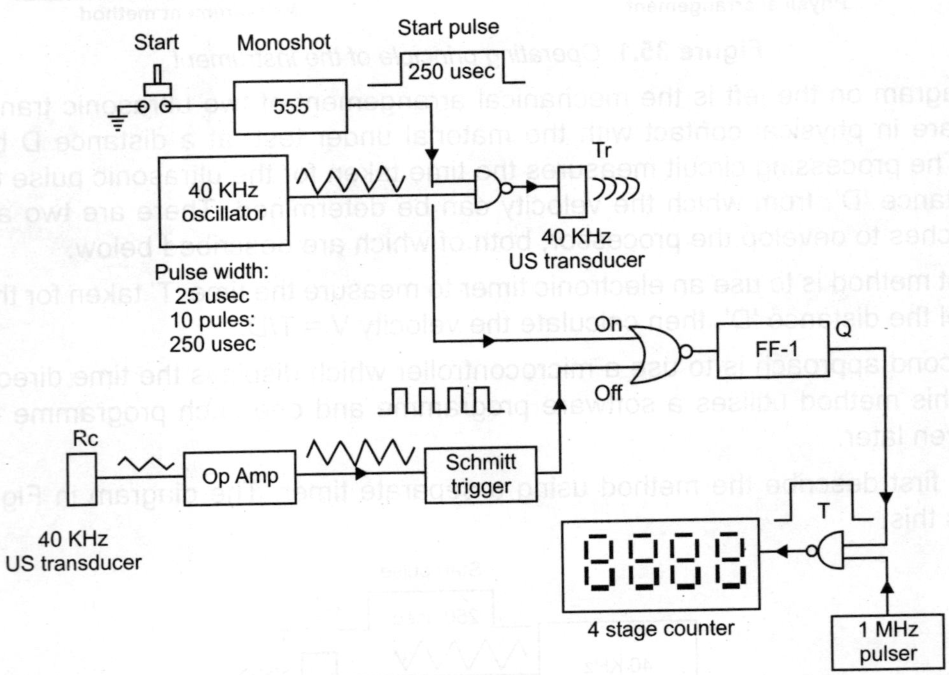

Figure 35.3 *Block diagram of complete system*

From Figure 35.3 we observe that the start pulse from the 555 monoshot is 250 microseconds wide. Since it is still high when the pulse from the receiver Schmitt trigger reaches the NOR gate at the input of the FF, it will upset the timing and indicate a false count.

There are two possible solutions here. The first is two use a larger sample for test purposes. In the case of iron, it should be at least 1.5 meters in length for reliable results. The second option is to reduce the duration of the start pulse from the monoshot to 125 microseconds. This will still inject five pulses from the transducer into the test piece.

Electronic Circuits

1. **Timing oscillator:** This is a 1 MHz pulser based on a quartz crystal for a stable frequency output. The circuit for this is given in Figure 35.4. It uses two gates from the IC 7400 package the other two gates being used elsewhere as shown in Figure 35.3 above.
   The 1 MHz pulses from the oscillator are connected to one of the inputs of the third gate in the same package. This gate will pass on the pulses to the counter when the other input receives the gating signal from the Q output of the Flip-Flop IC 7473.

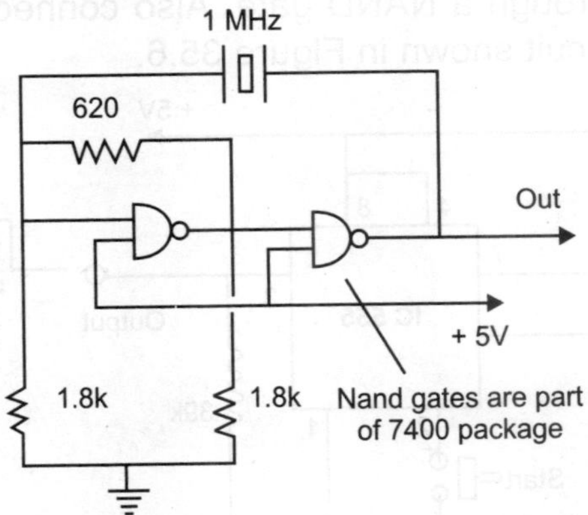

Figure 35.4 *Crystal controlled 1 MHz timing oscillator*

2. **Generation of ultrasonic pulses:** Ultrasonic transducers operating at 40 KHz are easily available, ranging in different physical sizes to suit individual requirements. The transducer recommended for this project is Murata type MA 40 EIR with an overall diameter of 18 mm. The active diameter is however 10 mm, the height of the transducer being 12 mm. This transducer can be wired directly onto a small PC board and placed over the area where the start pulse is to be injected on to the material under test.
   The transducer is powered by a 40 KHz oscillator based on the IC 555, as shown below in Figure 34.5.

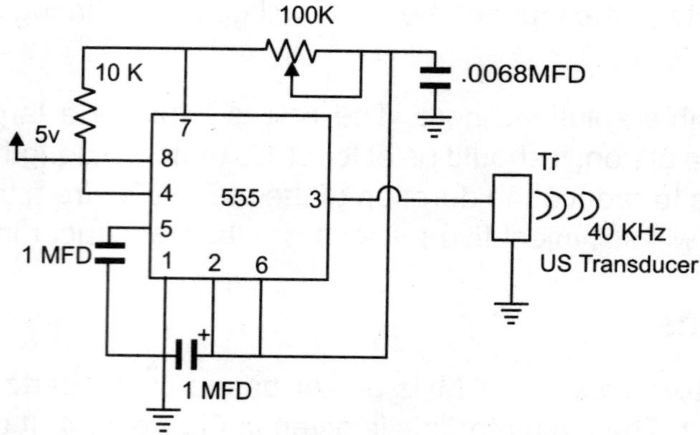

Figure 35.5 *40 KHz oscillator circuit for driving ultrasonic transducers*

The transducers are frequency selective, hence the oscillator circuit shown above has to be precisely set for 40 KHz by monitoring pin 3 on an oscilloscope. The frequency can be adjusted by manipulating the 100K pot shown in the circuit. Once set, the pot should not be disturbed.

It will be observed from Figure 35.5 that the 40 KHZ signal is not connected to the transducer directly, but through a NAND gate. Also connected to the gate is a single pulse generated by the circuit shown in Figure 35.6.

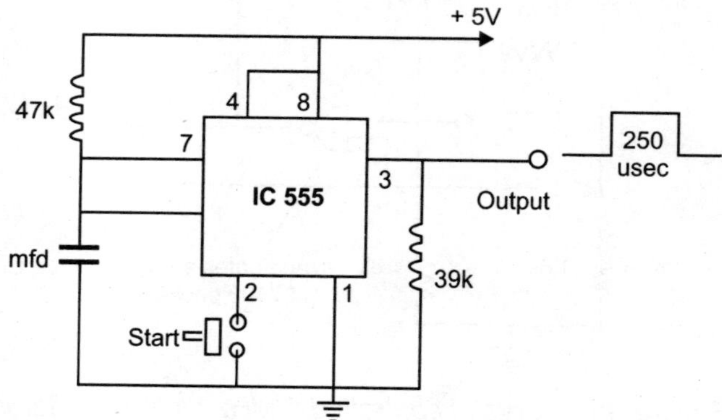

Figure 35.6 *Single pulse generator – monoshot – with 250 µsec output pulse*

This results in a single burst of 40 KHz signal injected into the transducer for a period of 250 µsec, during which interval only ten pulses will pass on to the transducer. This pulse is initiated by operated the press switch for a brief moment.

3. **Ultrasonic receiver and processing circuit:** A second transducer placed over the sample at a known distance receives the ultrasonic pulse train that travels through the material and converts this into an electrical signal of the same frequency. This is amplified by a two stage IC 741 opamp and fed to the Schmitt trigger IC 7413. This generates a square pulse which passes through NOR gate IC 7402 and acts as a clock pulse to toggle the FF IC type 7473. This FF should be reset using the Rest switch – normal position is + - which forces the Q output from the FF to go low.

The START pulse from the monoshot 555 is also connected to one input of the 7402 NOR gate. This toggles the FF Q output to go high. Since this is connected to the NAND gate which controls the 1 MHz signal to the counter, the pulses will pass through till the Q of the FF goes low again when the pulse from the Schmiit trigger IC 7413 toggles the FF again – through the NOR gate. The Q output of the FF goes low and the NAND gate block further counts from reaching the counter.

It will be observed that one of the NAND gates in the 7400 controls the pulses from the 1 MHz timing oscillator to the digital counter. Two gates in the same package form a part of the 1 MHz crystal oscillator.

The ultrasonic receiver circuit shown in Figure 35.8

The IC 7400 NAND gate pin-out is given below. Two of these form the crystal oscillator, while the other two perform the gating operations in the other parts of the circuit.

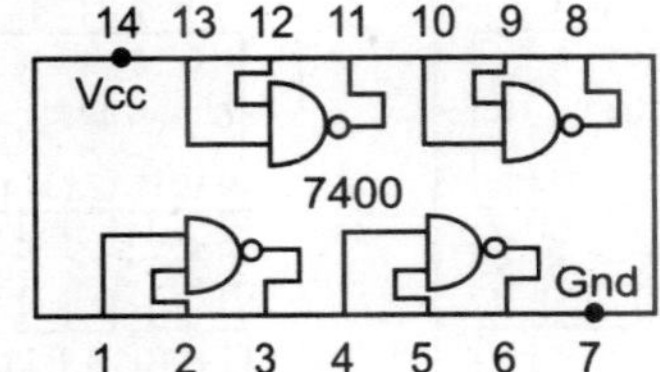

Figure 35.7 *IC 7400 NAND Gate*

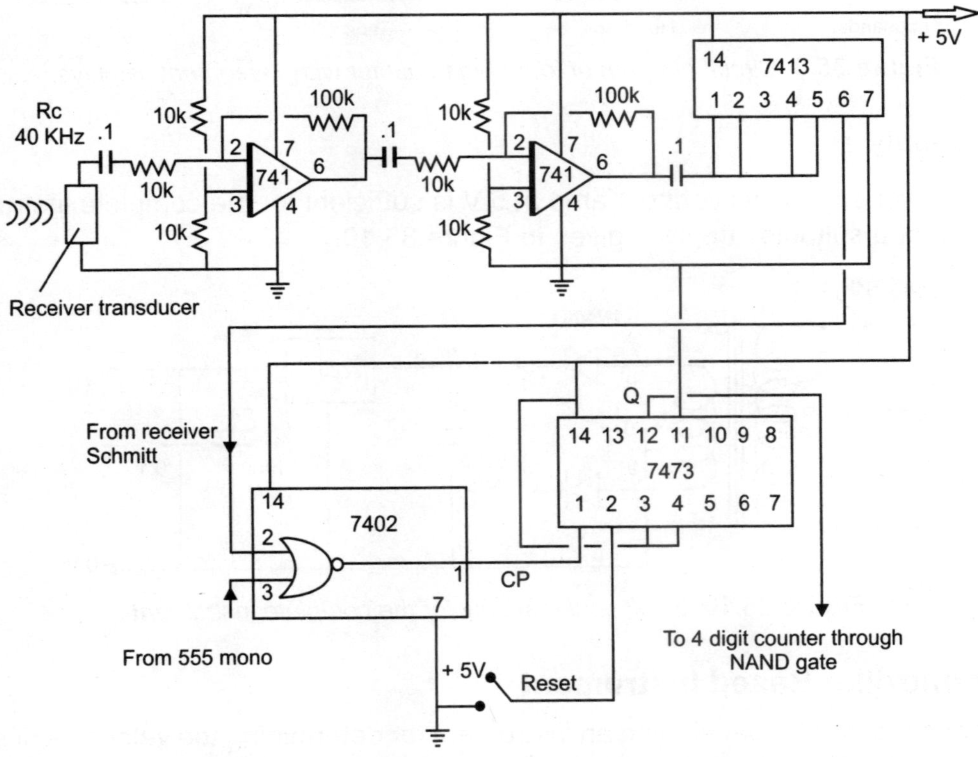

Figure 35.8 *Circuit showing the ultrasonic receiver, FF and gate*

The output from the control gate which allows pulses from the 1 MHz oscillator, is connected to a four-digit decimal counter which has a capacity of 9999 counts. This means that time durations up to 9999 microseconds can be displayed. The circuit of the counter-display is given in Figure 35.9.

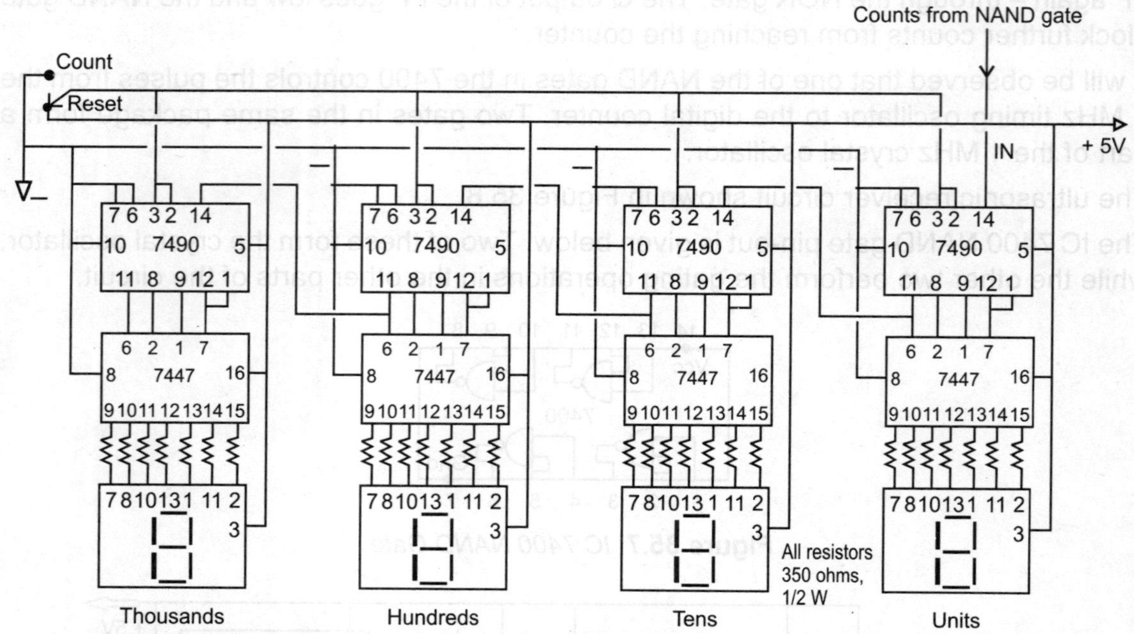

Figure 35.9 *Circuit diagram of four stage counter with 7-segment displays.*

Power Supply

A single power supply delivering 1 amp at 5 V is sufficient for the complete instrument. The circuit of a suitable supply is given in Figure 35.10.

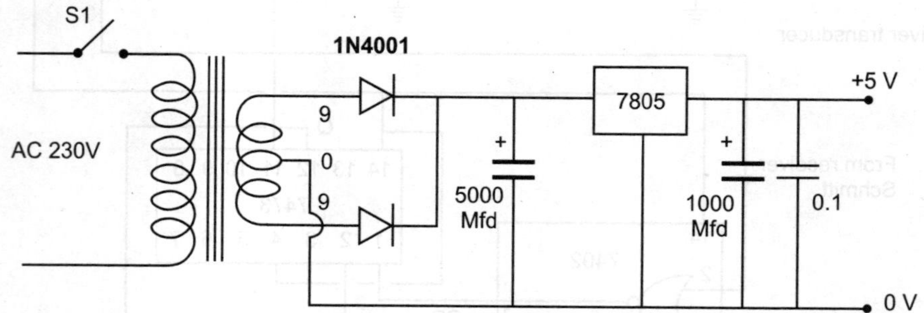

Figure 35.10 *5 volt power supply for the complete instrument*

Microcontroller Based Instrument

The second method for developing an instrument for determining the velocities of sound through solid bodies is based on the use of a microcontroller. This is described below:

The software for this project is written in 'C' language and is provided below.

A block diagram shown in Figure 35.11 below shows how a microcontroller can be used for measuring the velocity of sound through solid bodies where the test specimens are of short lengths.

The transmitter, the receiver and the crystal oscillator are identical to those described earlier.

The outputs of the transmitter start pulse and the receiver are connected to microcontroller type 8051, which displays the result on a LCD. Since the on and the off inputs at pins 12, and 13 are active low, the positive pulses from the Schmitt trigger of the receiver and the 555 monoshot have to be inverted. This is achieved through two of the three unused gates in the 7400 package.

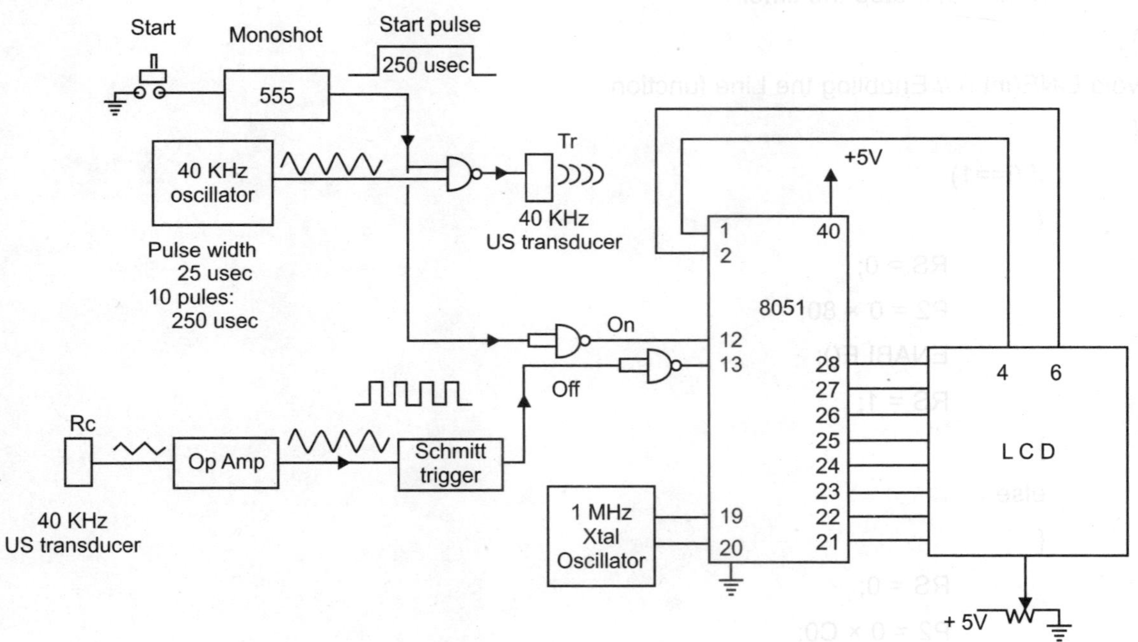

Figure 35.11 *Block diagram of microcontroller based instrument*

Software in 'C'

```
#include <REG52.H>
sbit RS = P1^0;
sbit EN = P1^1;
sbit interrupt1 = P3^3;  // External interrupt1
sbit interrupt0 = P3^2;  // External interrupt0
sbit IN =P3^4;           // input bit for external clock
```

```
void delay(int);
void INIT_LCD(void);
void ENABLE(void);
void LINE(int);
void ONTIME(void);
void timer0(void) interrupt 1 {
TH0 = 0 × C4;           //-60 ;COUNTING 60 PULSES
TR0 = 1;                // start the timer
}
void timer1(void) interrupt 3 {
        TR0 = 0; // stop the timer
}
void LINE(int i) // Enabling the Line function
{
        if (i==1)
        {
                RS = 0;
                P2 = 0 × 80;
                ENABLE();
                RS = 1;
        }
        else
        {
                RS = 0;
                P2 = 0 × C0;
                ENABLE();
                RS = 1;
        }
}
void delay()
{
int i,j;
        for (j = 0; j < 10; j++)
        {
```

```
                for (i=0;i<100;i++);
        }
}
void ENABLE(void) // Invoking the Enable function for the LCD.
{
        EN = 1;
        delay();
        //Exploring the Capabilities of On-Chip Resources Programming 65
        EN = 0;
        delay();
}
void INIT_LCD(void) // Initialization of the LCD by giving the proper commands
{
        RS = 0;
        EN = 0;
        P2 = 0 × 38; // 2 lines and 5?7 matrix LCD
        ENABLE();
        ENABLE();
        ENABLE();
        ENABLE();
        P2 = 0 × 06; // Shift cursor to left
        ENABLE();
        P2 = 0 × 0E; // Display ON, Cursor Blinking
        ENABLE();
        P2 = 0 × 01; // Clear display Screen
        ENABLE();
}
void ONTIME() // Invoking the ONTIME function
{
                TMOD=0x06; //000110B ;INITIALIZE COUNTER0,MODE2,C/T=1
        IE = 0 × 05; //0000101B INT0 & INT1 ENBABLED
  INIT_LCD();   //Initialize lcd
  LINE(1);
                        RS=1;           //data rs=1
```

```
P2 = 'S';
ENABLE();
RS = 0;
                    RS=1;
P2 = 'P';
ENABLE();
RS = 0;
                    RS=1;
P2 = 'E';
ENABLE();
RS = 0;
                    RS=1;
P2 = 'E';
ENABLE();
RS = 0;
                    RS = 1;
P2 = 'D';
ENABLE();
                    RS = 0;
                    RS = 1;
P2 = '(';
ENABLE();
RS = 0;
                    RS = 1;
P2 = 'm';
ENABLE();
RS = 0;
                    RS = 1;
P2 = '/';
ENABLE();
RS = 0;
                    RS = 1;
P2 = 's';
ENABLE();
```

```
RS = 0;
                    RS = 1;
P2 = 'e';
ENABLE();
RS = 0;
                    RS = 1;
P2 = 'c';
ENABLE();
                    RS = 0;
                    RS = 1;
P2 = ')';
ENABLE();
                    RS = 0;
                    RS = 1;
P2 = ':';
ENABLE();
RS = 0;
                    IN = 1;
}
void main (void){
        int i;
        int unit, tens, hundred;
                    ONTIME(); //init lcd ,int0 int1, timer0, display SPEED (m/sec):
                    while(1)
                    {
                    RS = 0;
                    EN = 0;
                    P2 = 0 × 0C; // display on cursor off
                    ENABLE();
                                        //setb tcon.0
                                        //setb tcon.2
                    IN = 1;             //make t0 as input
                    i = TL0;
                    RS = 0;
```

```
            P2 = 0 × 8D; // display on cursor off
            ENABLE();
            unit = (i%10);
            tens = (i/10)%10;
            hundred = (i/100)%10;
            unit = unit + 0 × 30; // Sending the ASCII data to the LCD
            RS = 1;
            P2 = unit;
            ENABLE();
            RS = 0;
            tens = tens + 0 × 30; // Sending the ASCII data to the LCD
            RS = 1;
            P2 = tens;
            ENABLE();
            RS = 0;
            hundred = hundred + 0 × 30; // Sending the ASCII data to the LCD
            RS = 1;
            P2 = hundred;
            ENABLE();
            RS = 0;
    }
}
```

END

Salt Content in Foods-Measurement

36

Salt is a mineral substance composed primarily of sodium chloride (NaCl), a chemical compound belonging to the larger class of ionic salts; salt in its natural form as a crystalline mineral is known as rock salt or halite. Salt is essential to the health of people and animals and is used universally as a seasoning. It is used in cooking, is added to manufactured foodstuffs and is often present on the table at mealtimes for individuals to sprinkle on their own food. Fortified salt consists of table salt mixed with a minute amount of potassium iodide, sodium iodide or sodium iodate.

Role of Salt in Human body

Table salt is made up of ~ 40% sodium by weight, so a 6 g serving (1 teaspoon) contains about 2,300 mg of sodium. Sodium serves a useful purpose in the human body: it helps nerves and muscles to function correctly, and it is one of the factors involved in the regulation of water content (fluid balance).

Recommended Intake of Sodium

Most of the sodium in the Western diet comes from salt. The habitual salt intake in many Western countries is about 10 g per day, and it is higher than that in many countries in Eastern Europe and Asia. The average intake of salt in many developed countries including India is above 9 g., as against the WHO recommended (1985) norm of 5 g, i.e., adults should consume less than 2,000 mg of sodium (5 g of salt) per day with some advocating for less than 1,200 mg of sodium (3 g of salt) per day.

Health Effects of Excess Salt Consumption

Even though research on the association between excess salt consumption and various health issues is limited, existing studies indicate that some of the long term health effects include stroke and cardiovascular disease, high blood pressure/ hypertension, left ventricular hypertrophy (cardiac enlargement), oedema and stomach cancer. In India salt caused high blood pressure or hypertension which was responsible for 57% of deaths due to stroke and 24% of deaths caused by heart attack. The majority of salt consumed is hidden in processed foods.

According to the World HealthOrganisation, 62% of all strokes and 49% of heart disease events can be attributed to high blood pressure. Sodium makes blood vessels less able to expand and contract and may toughen heart cells. Research on salt consumption suggests that reducing dietary salt intake to 3 grams per day (1,200 mg of sodium per day) could reduce the annual number of heart disease by 60,000, stroke by 32,000 and myocardial infarction by 54,000. It could also reduce annual number of deaths by any cause by 44,000. Also a 3 g per day reduction can bring down blood pressure by 2.5/1.4 mm Hg, and a 6 g per day reduction by 5/2.8 mm Hg. Conversely an increase of 5 g in daily salt intake was found to be associated with a 23% higher risk of stroke and 17% risk of cardiovascular disease. A low salt diet results in a greater improvement in blood pressure in those with hypertension than in those without.

Therefore, to reiterate the importance of the advantages of reduced salt intake, India can reduce incidents of stroke by 25% and heart attacks by 10% by cutting down on salt consumption.

While instruments to measure the amount of salt in foods are quite expensive, a device to do this can be fabricated by students with a basic knowledge of electronic instrumentation. Under this project we describe one such instrument which requires commonly available components and hardware.

The instrument has four parts. A DC power supply provides 9 V for the Wien bridge oscillator which generates sine waves at around 400 hertz. The oscillator output is connected across the conductance bridge. A probe from the conductance bridge will be used to measure the salt content of various foods.See Figure 36.1.

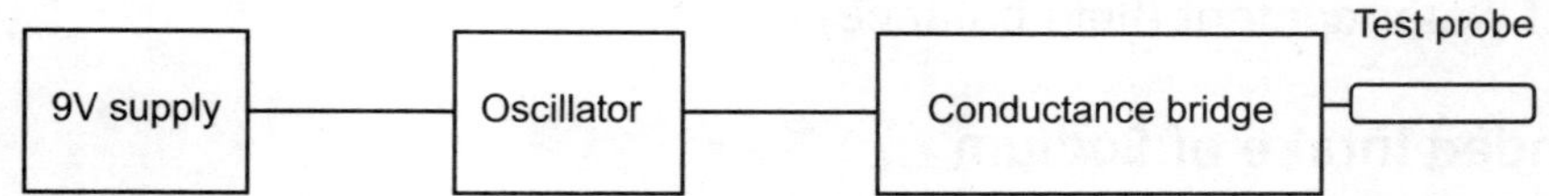

Figure 36.1 *Block diagram of the complete instrument*

The 9 V required for the Wien bridge oscillator will be provided by a standard supply, a circuit of which is provided in Figure 36.2 below:

It is a fairly straightforward circuit with IC 7809 providing a stabilised 9 V for the Wien bridge oscillator. Since the transformer requirement is only 250 mA, the power supply, the Wien bridge oscillator and the conductance bridge can all be built on a single chassis, with only the probe as a separate unit.

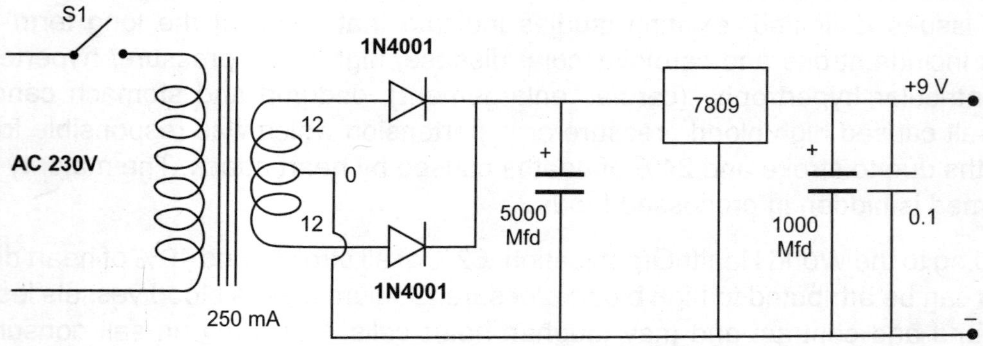

Figure 36.2 *Circuit diagram of power supply for Wien bridge oscillator*

Since salt conducts electrical current, the amount of current flow is directly proportional to the salt content. In foods, monitoring the electric current gives an indication of the quantity of salt in it. Due to polarisation effects, DC type ohmmeters cannot be used for this purpose. AC conductance meters are generally used for monitoring the salt level in food products.For this project we use a Wien bridge oscillator to provide about 400 Hz AC for the conductance bridge. This circuit is shown in Figure 36.3.

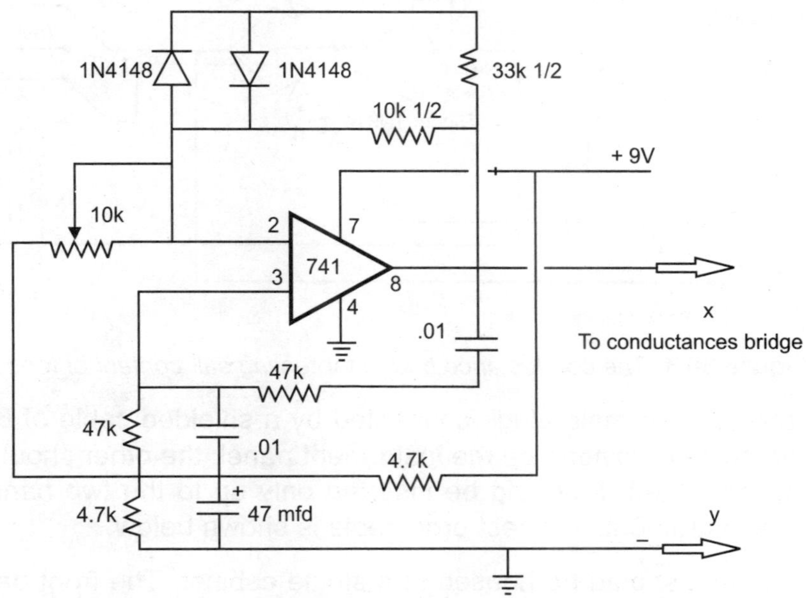

Figure 36.3 *Sine wave oscillator for the conductance bridge*

The oscillator is built around the 741 op amp, with frequency determining components being the 47 k resistances and the .01 Mfd capacitors. Output of the oscillator is provided at points marked x, y, and connect to the conductance bridge shown below.

Conductance Bridge

The bridge uses 400 Hz sine wave as the source. The out-of balance AC signal is rectified by the diode capacitor network and fed to a dc mA meter with a full scale deflection of 1 mA.

In the circuit shown in Figure36.4, the 5k pot forms a part of the bridge arms, with the wiper connected as shown. This enables the calibration of the instrument as detailed below. The transformer connecting the diode has a primary winding of 20,000 ohms and a secondary of 1,000 ohms.

In use, the probe is inserted into the food to be tested. This is a standard stereo plug commonly used in music systems. The two connections to the probe will result in the resistance of the food sample to be placed in parallel with the 1k resistance in the bridge arm. This unbalances the bridge and a current flows into the 20k primary winding of the transformer. The rectified current flows through the meter which gives an indication of the salt content in the food.

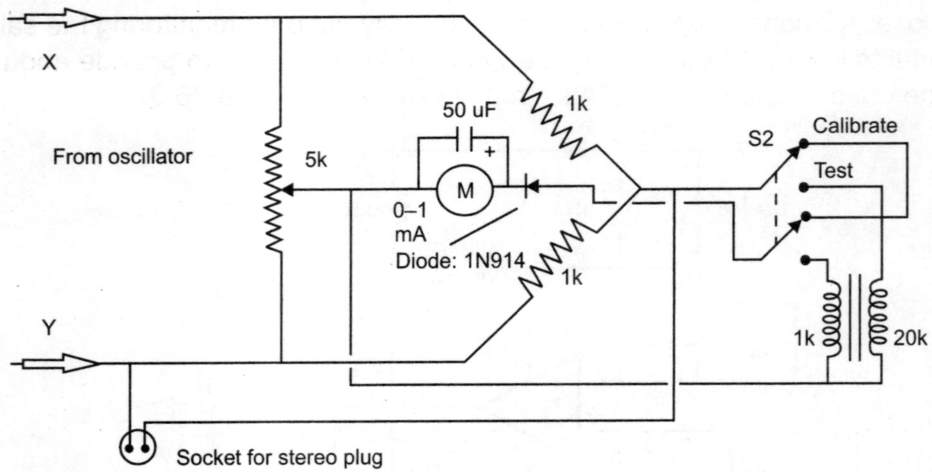

Figure 36.4 *The conductance bridge for testing salt content of foods*

The probe cable has two male plugs connected by a shielded cable of 600 mm. One plug fits into the socket provided on the instrument panel, the other should be inserted into the food to be tested. It should be inserted only up to the two bands which are connected to the instrument. The test prod cable is shown below.

The entire instrument should be housed in a single cabinet. The front panel will have the switches, the meter, the calibration pot and the probe socket.

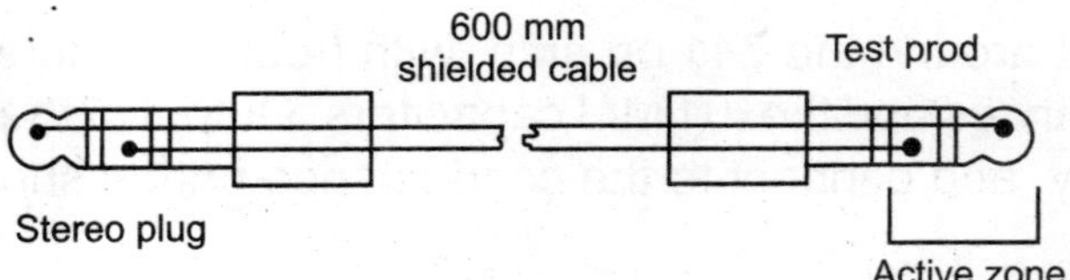

Figure 36.5 *Test prod cable with two plugs*

The figure below is a suggestive layout of the front panel.

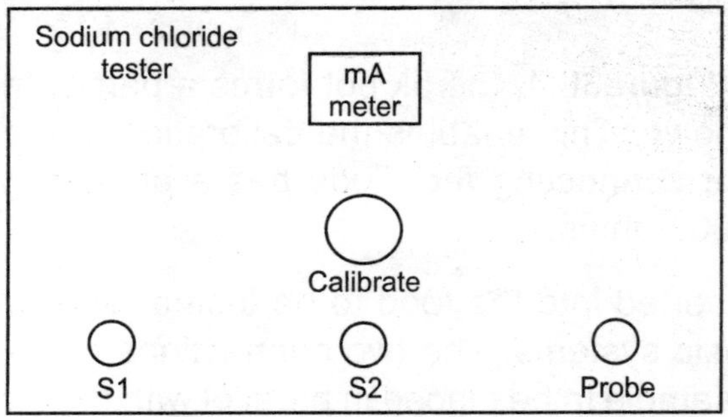

Figure 36.6 *Suggested layout of front panel*

The instrument can be calibrated by noting readings of solutions with known salt content. Tap water shows a resistance of 20,000 ohms when the probe is connected to an ordinary ohm-meter. With the switch S2 in calibrate position, adjust the 5 k pot till the meter needle points to zero position.

In 100 ml of water, dissolve 0.5 grams of salt and measure the conductivity of the solution. Then add another 0.5 grams of salt and measure the conductivity again. Repeat this test another three times, noting the conductivity each time. Draw a curve to represent the meter reading vs the salt content. Before each test, the probe should be wiped clean.

37 Van De Graaf Generator of Testing Insulator

A high voltage generator of the Van De Graaf type, yielding 50 K.V for testing insulators is described under this project.

Such generators were originally developed for use with nuclear particle accelerators. It has also found applications wherever very high voltages – with low current – are required. Such high voltage generators operate in open air as well as in compressed air environments – by which the voltage outputs can be increased with the same physical dimensions.

Operating Principle

A high voltage terminal – generally a hollow metallic sphere – is mounted on an insulating column. A rubber belt runs from the bottom of the insulating column to the inner side of the terminal – over a set of pulleys.

Electric charges are sprayed on to the lower part of the belt. These are carried into the terminal and deposited there. The voltage of the terminal rises due to this continuous accumulation of charge. The terminal voltage depends on its physical parameters, the environment in which the device operates, the insulating column etc.

A diagram showing one such generator developed at the Hyderabad Science Society and which yielded 500 kiloV, is shown in Figure 37.1.

Under this project, a small Van De Graaf generator is described which can generate up to 50,000 V in open, dry, air. An overall view of this generator is shown in Figure 37.2.

Components of the Generator

The complete system can be divided into three main parts:

1. A lower platform with a motor-pulley and a high voltage supply to spray electrical charges onto a fast moving rubber belt, which moves vertically.
2. An insulating column which supports a high voltage terminal and isolates it from the ground, and through which the rubber belt carrying the charges travels.
3. A high voltage terminal - into which the belt enters - and collects the charges from the belt, increasing its potential very rapidly.

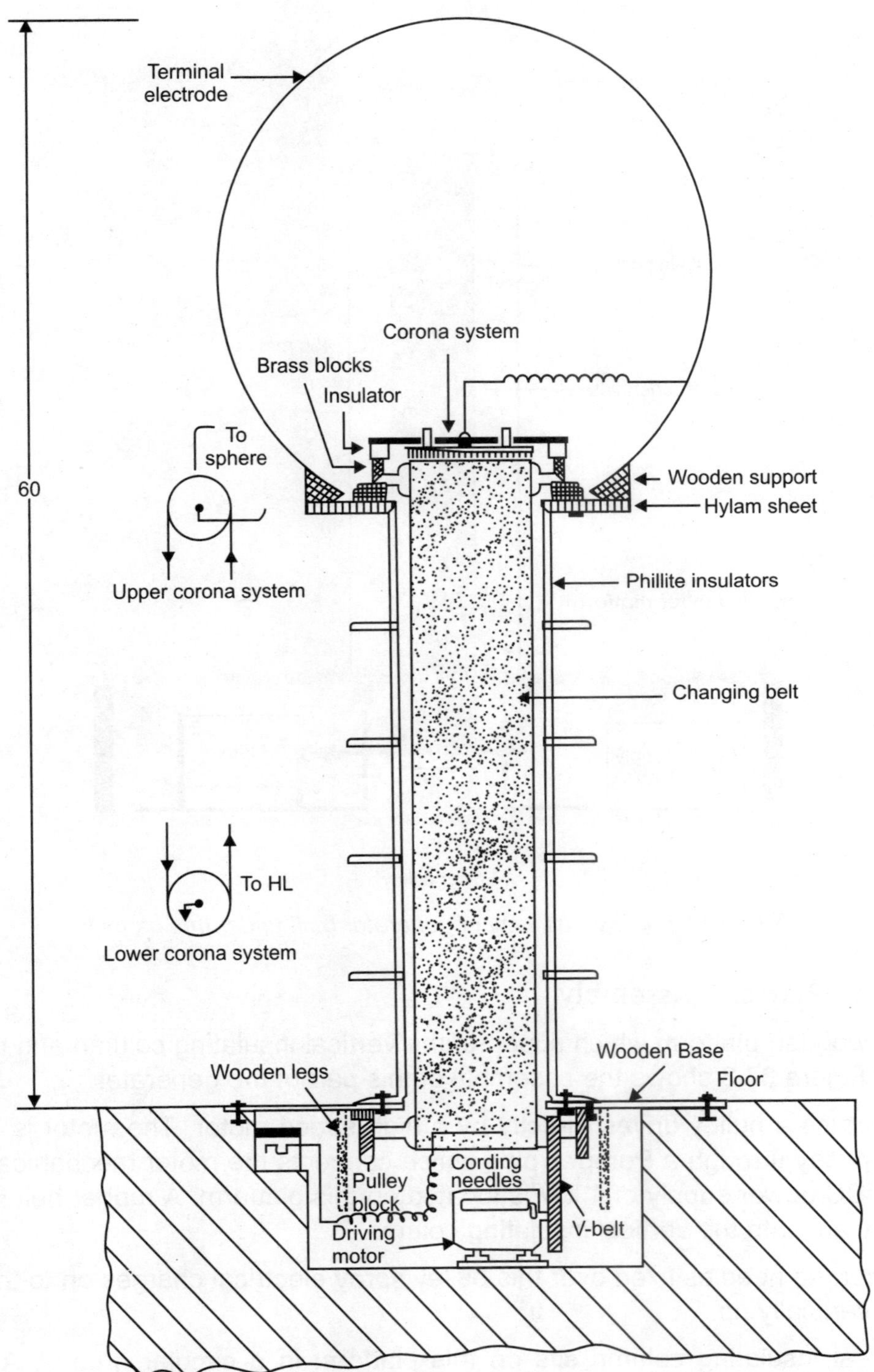

Figure 37.1 *Original 500 K V Van De Graaf Generator*

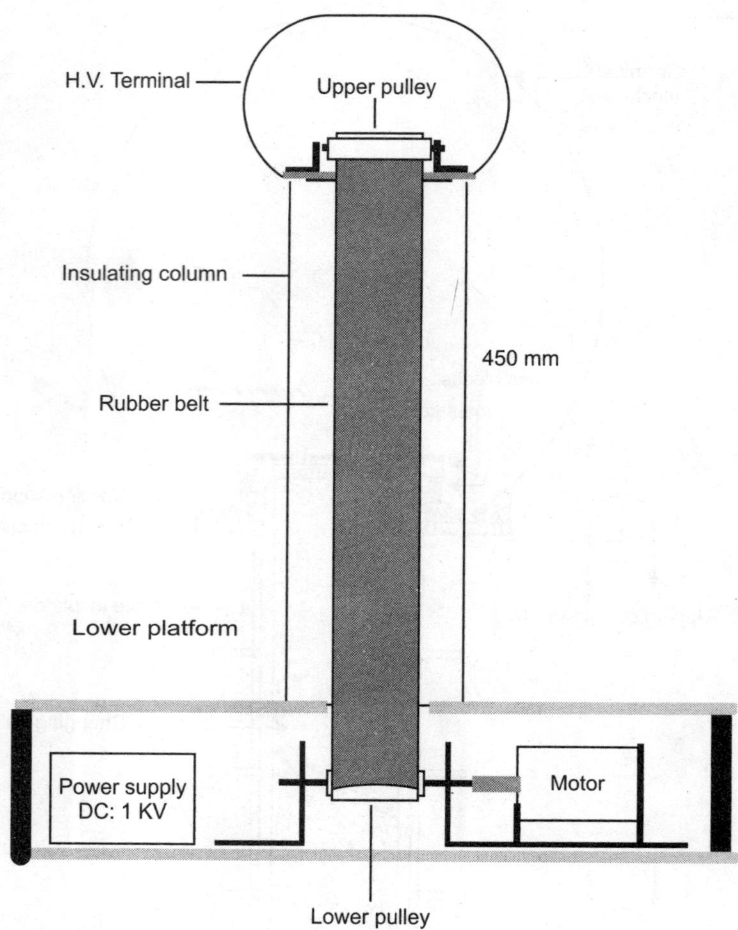

Figure 37.2 *Overall view of generator built under this project*

The Lower Platform Assembly

This is a wooden platform which supports the vertical insulating column and the upper terminal. Figure 37.3 shows the assembly of this part of the generator.

It incorporates a pulley driven directly by a high speed motor. The motor is insulated from the pulley through a Perspex pipe which connects the motor mechanically to the pulley. A DC power supply of 1 k V is located on this platform. A rubber belt runs over this pulley and into the vertical insulating column.

A set of corona needles fixed over this pulley spray electrical charges on to the belt as it moves vertically up.

The vertical insulating column sits on this platform in a circular groove. A wooden supporting collar is provided to hold the insulating column vertical and rigidly fixed over the platform. The bottom part of the platform is also a circular wooden base of the same dimensions as the top part. The top part of the platform is supported by three wooden legs which are screwed on to the top and the bottom bases, providing a gap of 300 mm in between.

The vertical insulating column sits on this platform in a circular groove.

A wooden supporting collar is provided to hold the insulating column vertical and rigidly fixed over the platform. The bottom part of the platform is also a circular wooden base of the same dimensions as the top part. The top part of the platform is supported by three wooden legs which are screwed on to the top and the bottom bases, providing a gap of 300 mm in between.

The vertical insulating column sits on this platform in a circular groove. A wooden supporting collar is provided to hold the insulating column vertical and rigidly fixed over the platform. The bottom part of the platform is also a circular wooden base of the same dimensions as the top part. The top part of the platform is supported by three wooden legs which are screwed on to the top and the bottom bases, providing a gap of 300 mm in between.

The vertical insulating column sits on this platform in a circular groove. A wooden supporting collar is provided to hold the insulating column vertical and rigidly fixed over the platform. The bottom part of the platform is also a circular wooden base of the same dimensions as the top part. The top part of the platform is supported by three wooden legs which are screwed on to the top and the bottom bases, providing a gap of 300 mm in between.

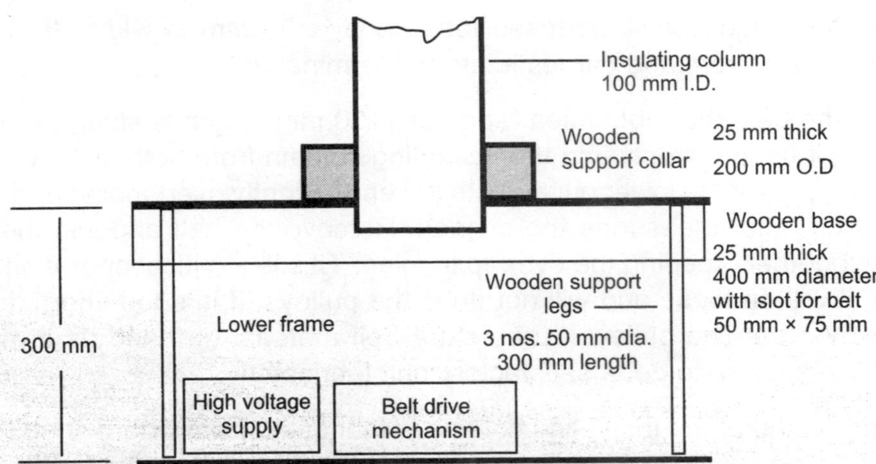

Figure 37.3 *Cross section of lower platform which supports the upper assembly*

The lower base supports the belt drive assembly and the high voltage power supply. The belt drive mechanism is shown in Figure 37.4 in coloured illustrations.

A set of sharp metallic points, known as corona needles, supported by an insulating strip, are located horizontally over the portion where the belt moves in the upward direction. The belt moves between these needles and the pulley, the needles maintaining a gap of approximately 2 mm from the belt. These needles are connected to the positive side of the high voltage supply, the negative side being connected to the pillow blocks supporting the pulley. Thus there is a voltage of 1,000 V between the corona needles and the pulley. This results in a steady transfer of charge which is intercepted by the belt as it moves upwards.

Insulating Column

The insulating column is a Perspex pipe with an external diameter of 100 mm and a wall thickness of 2 mm. It is 450 mm in length and carries the charging belt within it. The top part supports the high voltage terminal while the lower part sits on the lower platform. The column fits into a groove in the bottom base plate, as shown in Figure 37.4. in coloured illustrations on pages 177–184. It has a collar to provide extra support. The column is held in position by the belt which runs over the top and the bottom pulleys.

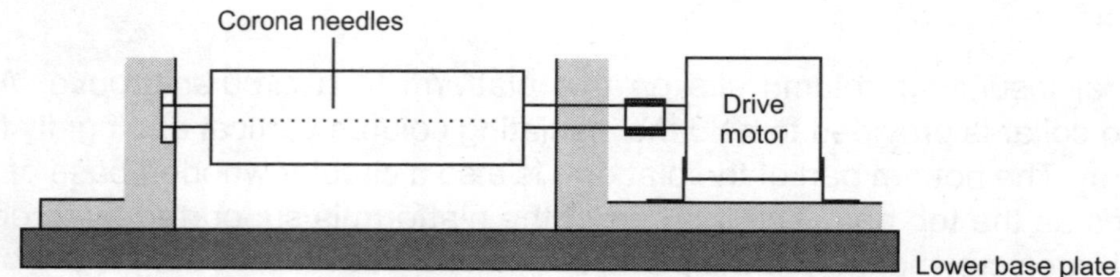

Figure 37.4 *Lower belt derive system*

The belt is fabricated from rubberised fabric and is 50 mm in width. It picks up the charges from the bottom and carries it into the terminal at the top.

To fabricate the belt, the rubberised fabric of 1100 mm length is slung over the upper pulley so that it hangs evenly into the insulating column from both sides of the pulley. The belt should cover the lower pulley, with the ends slightly overlapping and the excess material cut off, after measuring the overlap. Remove the belt and join the ends and paste them together, keeping the overlap in mind. This is a critical operation. If the belt is too long, it will be loose and will not drive the pulleys. If it is too short, it cannot be assembled over the two pulleys. Keep extra belt material on hand as it may require more than one attempt to get a satisfactory belt fabricated.

The platform on top of the insulating column is fabricated from hylam sheet. A wooden collar supports the insulating column; see Figure 37.5 in coloured illustrations on pages 177–184.

The upper pulley structure and the high voltage terminal sit on this hylam sheet.

The hylam sheet shown in Figure 37.5 has a centrally located slot of size 60 mm × 40 mm to accommodate the belt which enters the high voltage terminal. The upper and the lower platforms have grooves to accommodate the insulating column and are held in position by the tension of the belt.

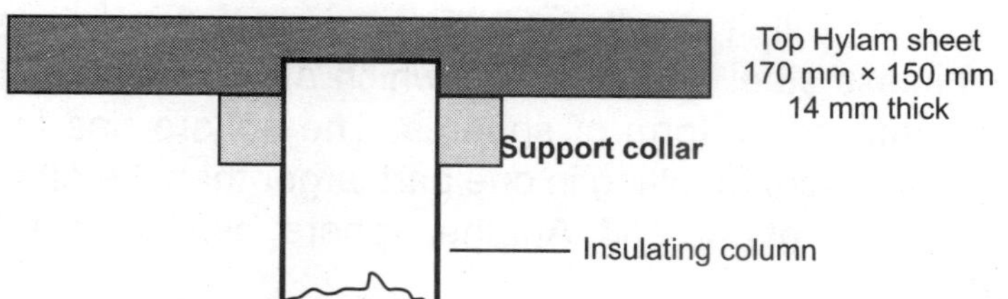

Figure 37.5 *Upper supporting platform insulating column*

The High Voltage Terminal Assembly

The upper platform accommodates the high voltage terminal as well as the upper pulley system located within the terminal.

The diagram in Figure 37.6 in coloured illustrations, shows the cross section of the upper pulley system. A set of corona needles located over the pulley collect the electric charges from the belt and deposit these on to the high voltage terminal.

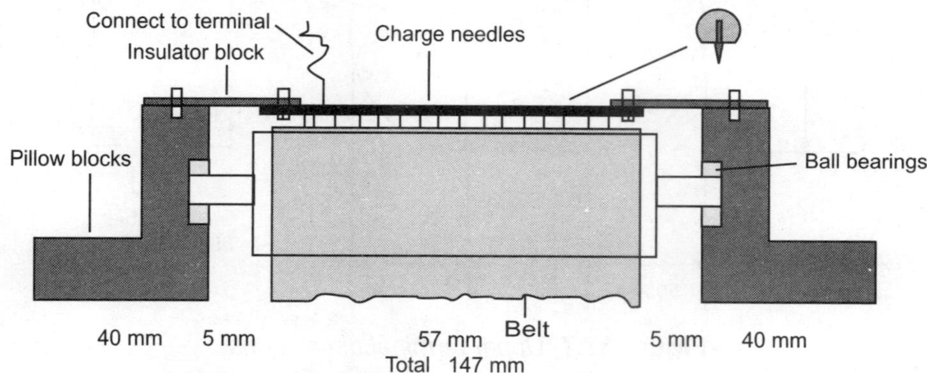

Figure 37.6 *Upper pulley assembly with charging needles*

As may be observed from Figure 37.6 in coloured illustrations on pages 177–184, the charging needles are insulated from the pillow blocks. They connect though a wire to the high voltage terminal.

The high voltage terminal is assembled from three sections made of stainless steel. Two of these are hemispherical end cups of 100 mm diameter. The third section is cylindrical in shape and accommodates the two hemispheres at its ends. Details of the terminal are given in Figure 37.7

The terminal is cylindrical in shape with the ends terminating in two hemispheres, one at each end.

Fabrication of the High Voltage Terminal

The terminal has two parts:

1. A cylindrical mid-section fabricated from stainless steel sheet as per the drawing shown in Figure 37.7. This cylinder has a slot through which the belt will pass into the terminal.

2. Two end cups which will fit into the cylinder and close the cylinder ends. The end cups are of stainless steel and of a size which are readily available in hardware shops. These come in the form of spheres. The sphere has to be cut at a little below the circumference, resulting in one part larger than the other. The larger part will fit into the cylinder at one end. Another sphere, similarly cut, fits into the other end of the cylinder.

The cylinder and the end cups should be brazed and the ends smoothed out. It is important that there should be no sharp edges or points on the terminal outer surface.

It is not necessary that the terminal should be of stainless steel. Aluminium or brass will also work well, provided that spheres of the right size are available easily.

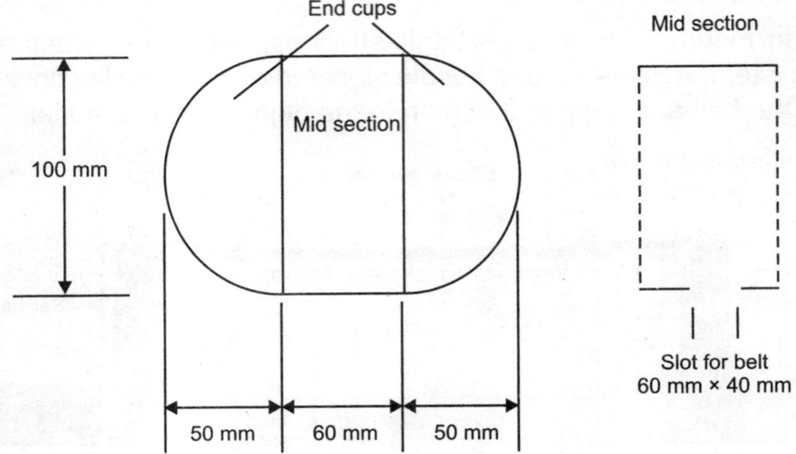

Figure 37.7 *Upper terminal dimensions*

The belt moving over the lower pulley is prayed with electric charges from a 1K V DC power supply. A suitable power supply is shown below.

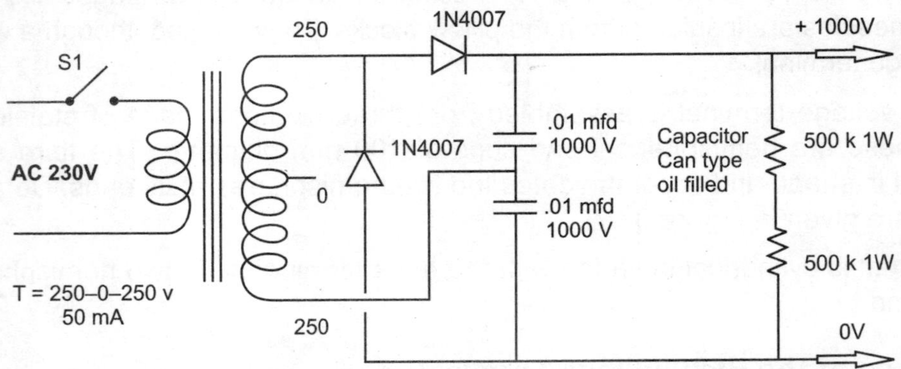

Figure 37.8 *High voltage power supply for charging the belt*

A standard 250 – 0 - 250 V transformer is used for the supply. The centre tap is ignored while the two ends of the secondary with output voltage of 500 V are connected in a voltage - doubling circuit consisting of the two diodes and the 0.01 mfd, oil filled capacitors, yielding 1000 V output. The two 500k resistors act as a bleeder to discharge the capacitors on switch-off.

Assembly of the Van De Graaf generator

1. The upper platform and the pulley assembly is prepared first, with the corona needles sitting on top of the pulley assembly. The belt should be slung over the upper pulley so that it will hang down into the vertical column.
2. The lower pulley system is assembled next.
3. The central insulating column is then inserted into the lower platform groove.
4. The upper platform is inserted onto the insulating column so that the column fits into its groove. The belt should go through the column and the lower pulley located between it.
5. The lower pulley, with the belt is fixed on to the lower platform.
6. Adjust the lower pulley so that the belt is now firmly lodged between the two pulleys.
7. Attach the drive motor to the lower pulley and take a trial at this stage to check if the belt drive system is operating satisfactorily. This completes the mechanical assembly of the system.
8. Fit the high voltage terminal over top platform and ensure that it is rigidly fixed. The wire from the corona needle should be connected to the high voltage terminal.
9. Connect the wire from the live side of the 1 k V power supply to the lower corona needles; the ground side of this supply connects to the lower pulley block.

Operation

1. Switch on the lower drive motor. When it starts running, switch on the 1 k V high voltage supply. The upper terminal will reach its full voltage capacity with a couple of minutes. This can be confirmed by using a spark gap consisting of a 100mm diameter, metallic sphere connected to the ground terminal. This sphere can be suspended over the high voltage terminal though an adjustable chord, so that the distance between the high voltage terminal and the grounded sphere can be varied. In dry weather, you should see a spark when the distance between the high voltage terminal and the sphere is 50 mm, if the voltage has reached 50 kv.
2. The lower high voltage supply should be switched off before stopping the belt drive system.

Note: The Van De Graaf works on static electricity. It will not operate if the ambient humidity is high or if there is a lot of dust on the belt, or if the distance of the corona needles from the pulleys it too large, or if the lower high voltage supply is not sufficient to spray charges on to the moving belt.

If the upper high voltage terminal is not smooth and has sharp corners, electric discharges may take place before the full voltage is reached.

❑❑❑

Coloured Illustrations

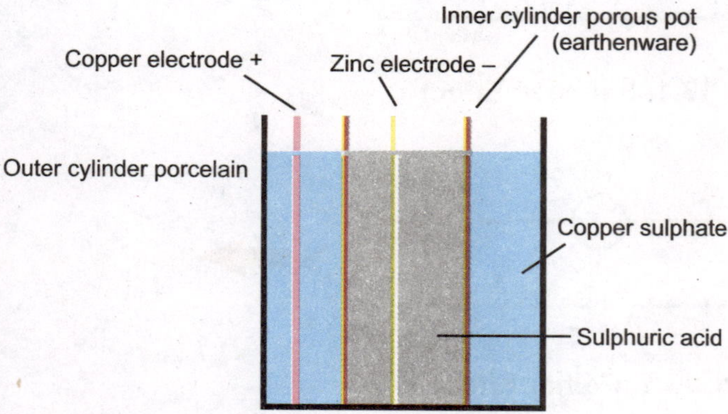

Fig. 3.1: Cross section of a Daniel Cell

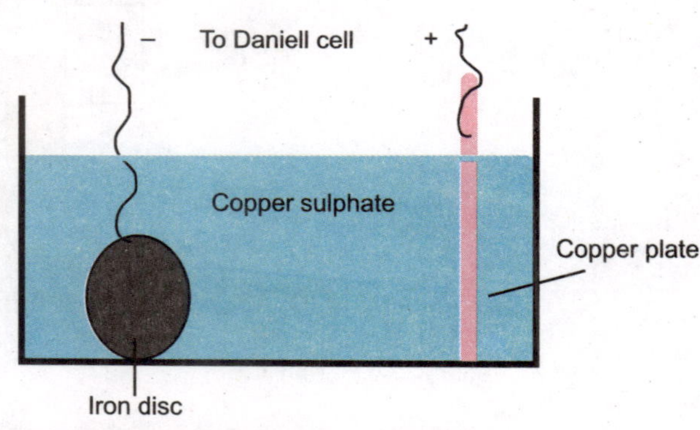

Electroplating bath: Copper gets deposited from the plate to the iron disc.

Fig. 3.2: Arrangement for copper plating

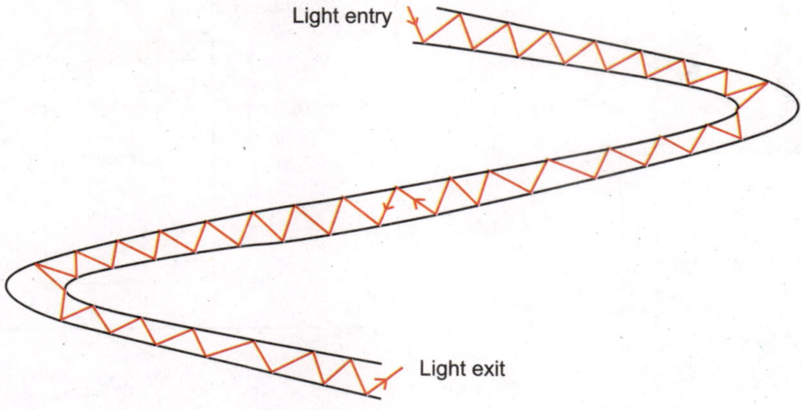

Fig. 7.1: Light travelling by total internal reflections in an optical fibre

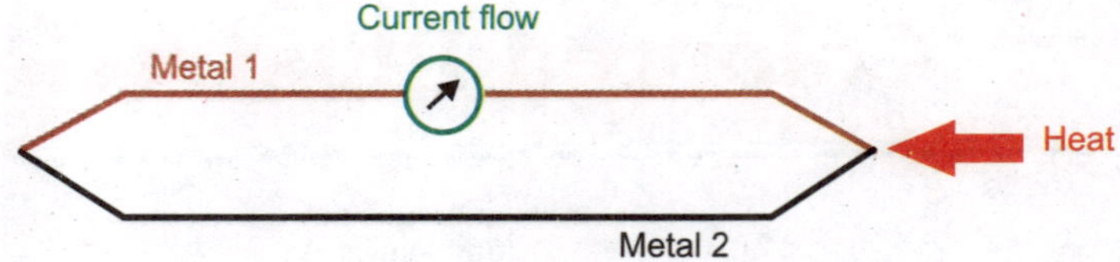

Fig. 13.1: Seebeck Effect

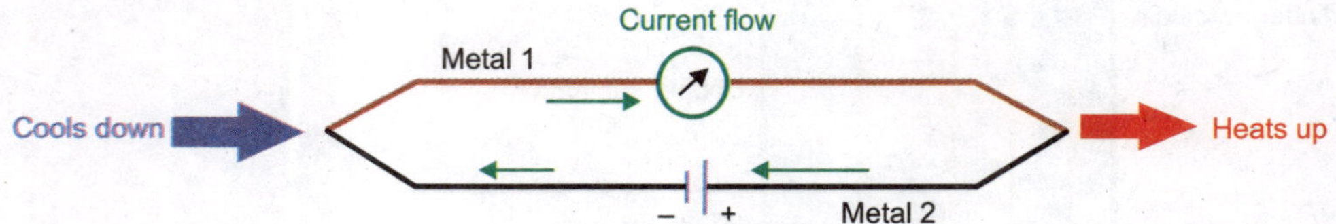

Fig. 13.2: Peltier Effect

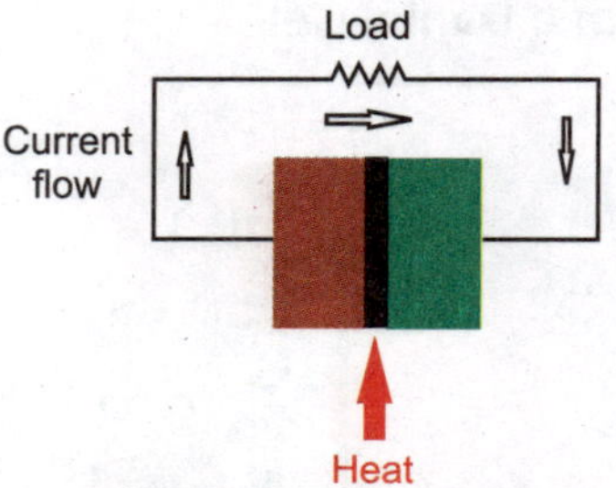

Fig. 13.3: Seebeck Effect in a p-n junction

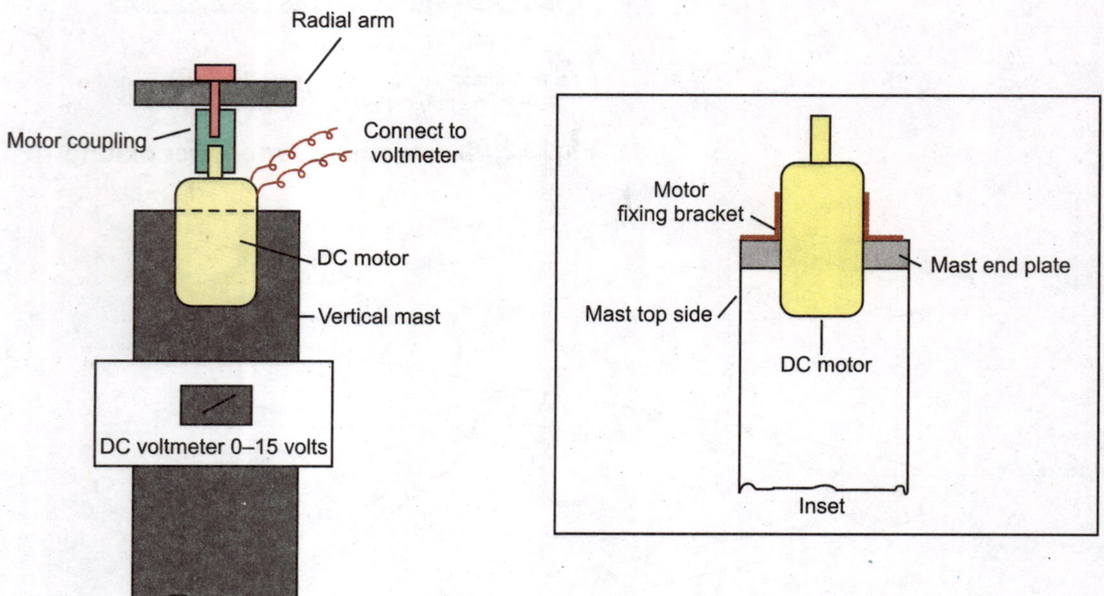

Fig. 22.2: Assembly of motor and voltmeter on mast

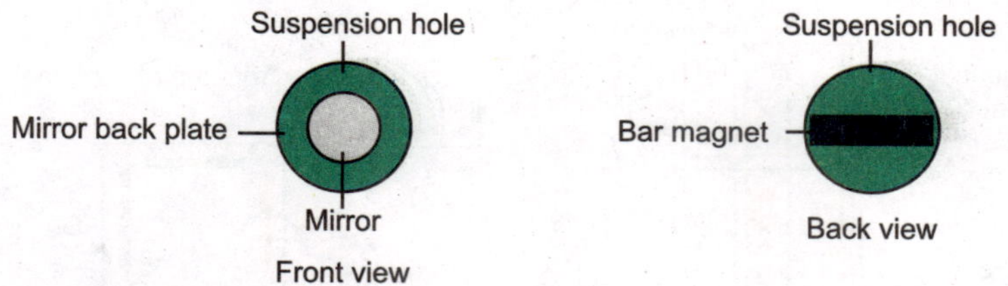

Fig. 24.3: Fabrication of the mirror-magnet unit

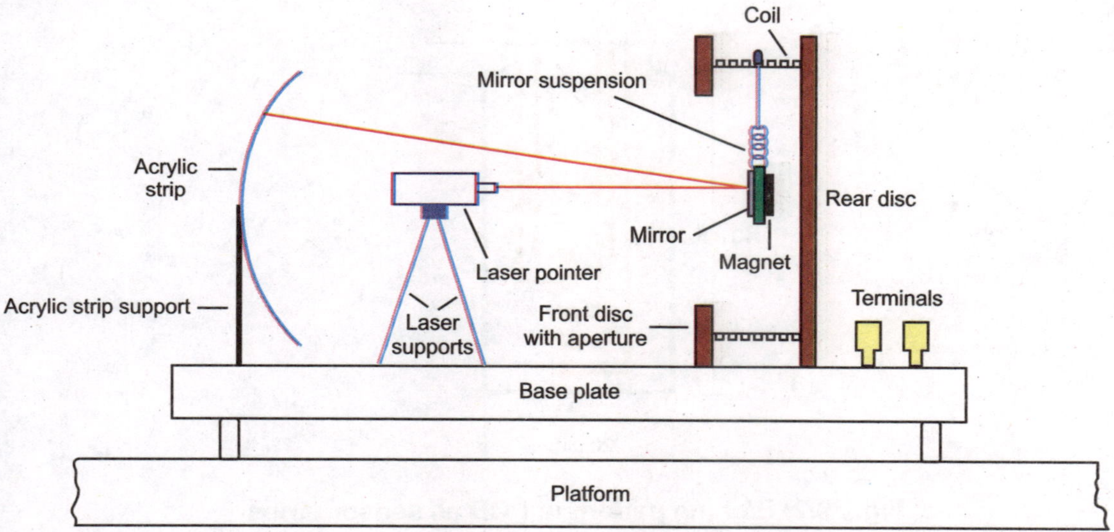

Fig. 24.6: Overall view of the galvanometer assembly

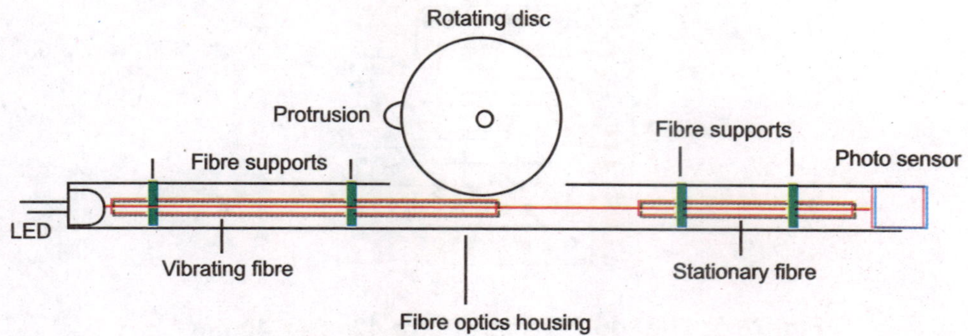

Fig. 27.5: Fibre optics assembly in U channel

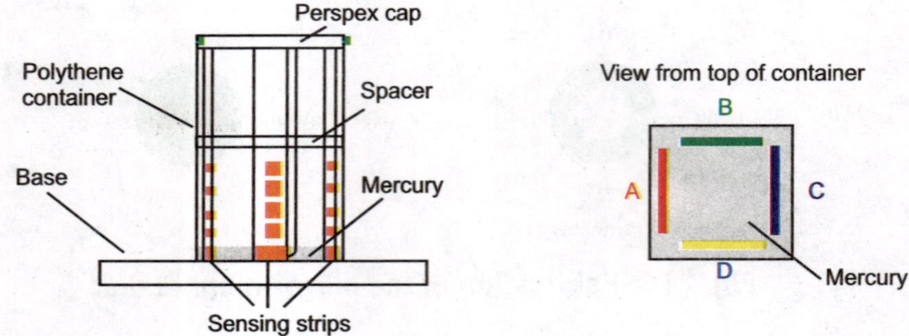

Fig. 29.1: Tilt Sensor assembly

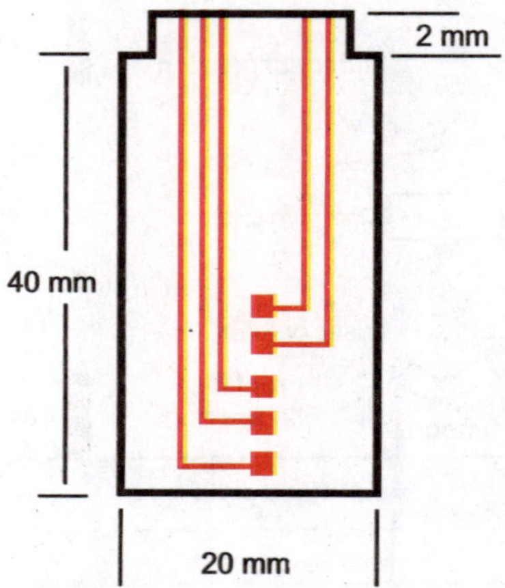

Fig. 29.2: Etching pattern of PCB as sensor strips

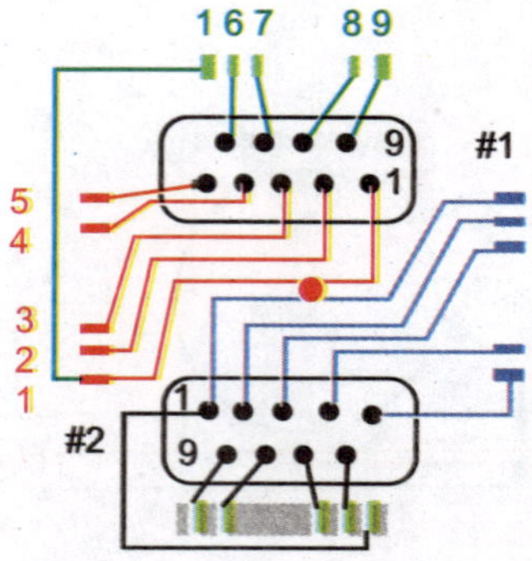

Fig. 29.5: The top PC board: size 40mm x 40mm Etching pattern shown above

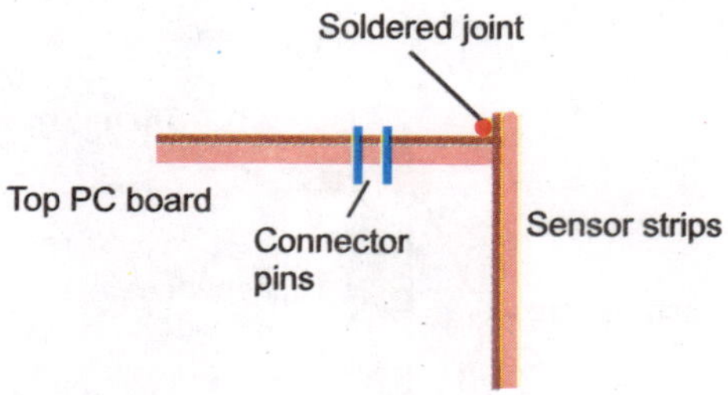

Fig. 29.6: Connecting the sensor strips to the top PCB

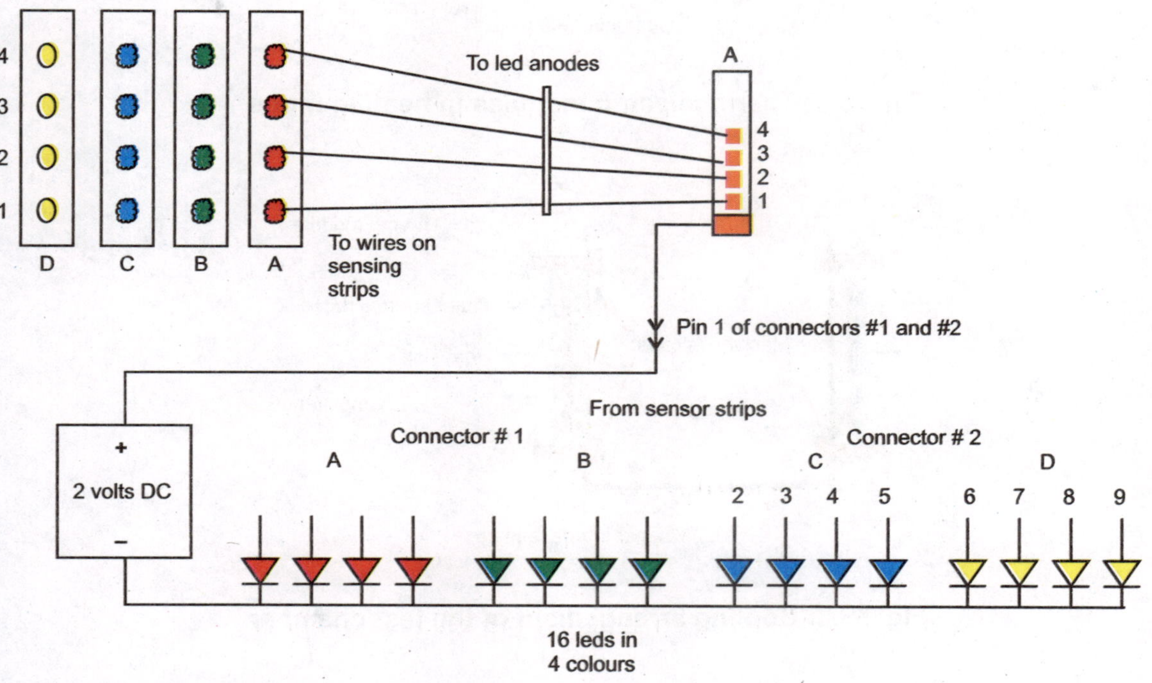

Fig. 29.7: Wiring diagram for the LEDs located on the front panel of the display unit

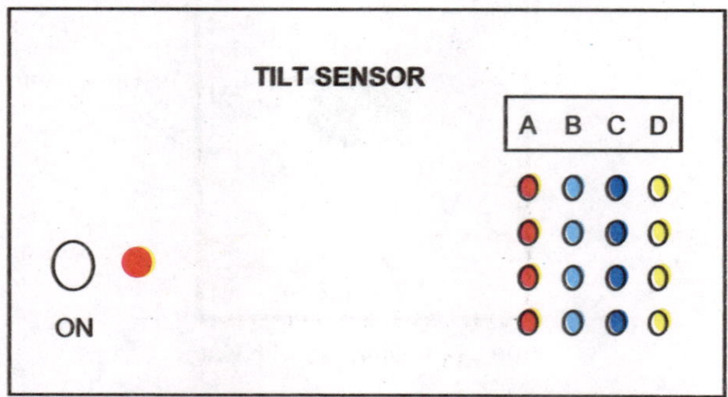

Fig. 29.9: Front view of the Tilt Sensor

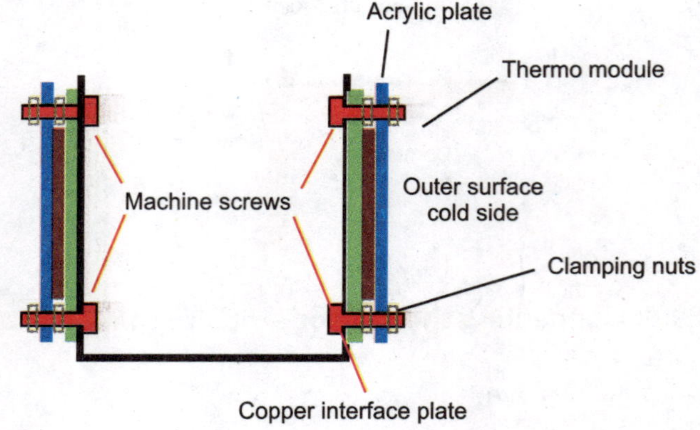

Fig. 33.2: Thermoelectric modules in heating mode

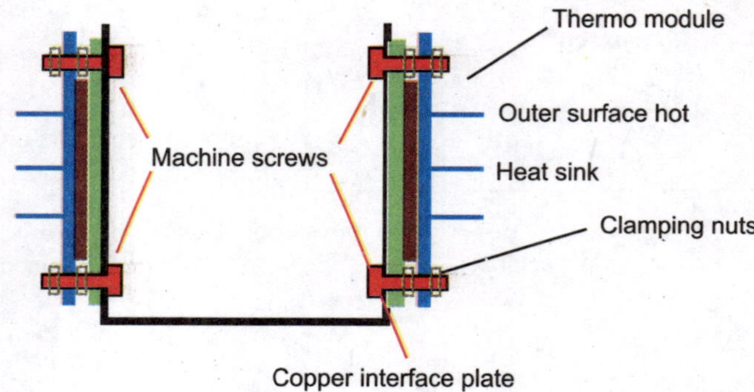

Fig. 33.3: Cooling arrangement of the test chamber

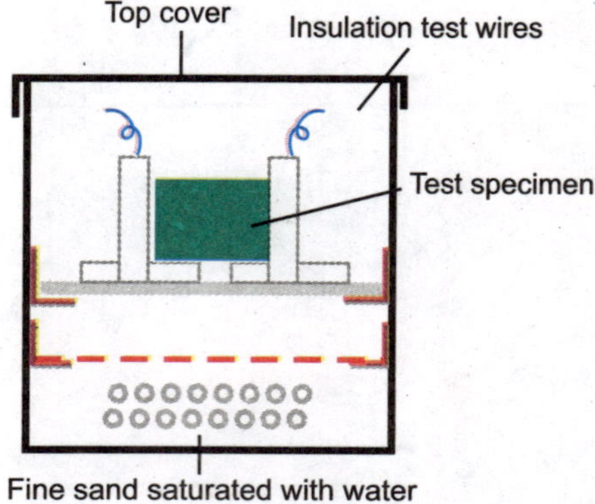

Fig. 33.4: Cross section of test chamber

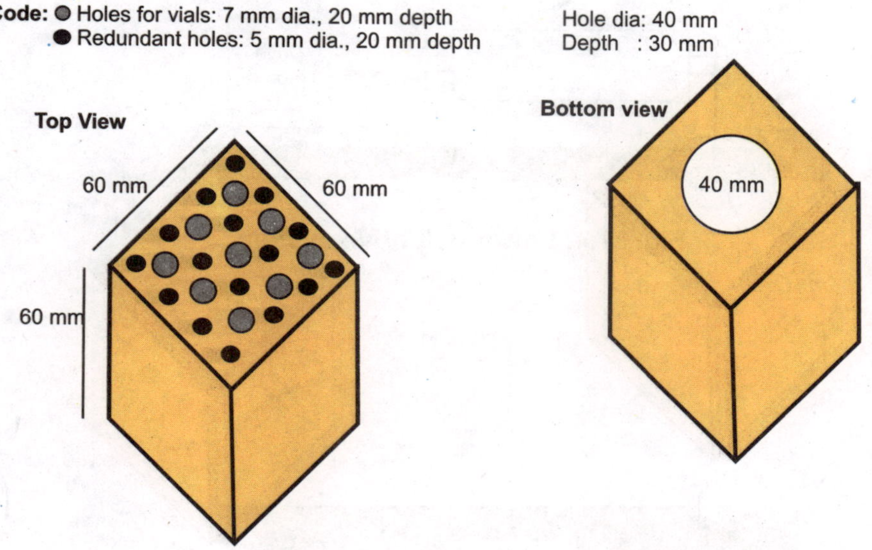

Fig. 34.1: Two views of metallic block for heating vials

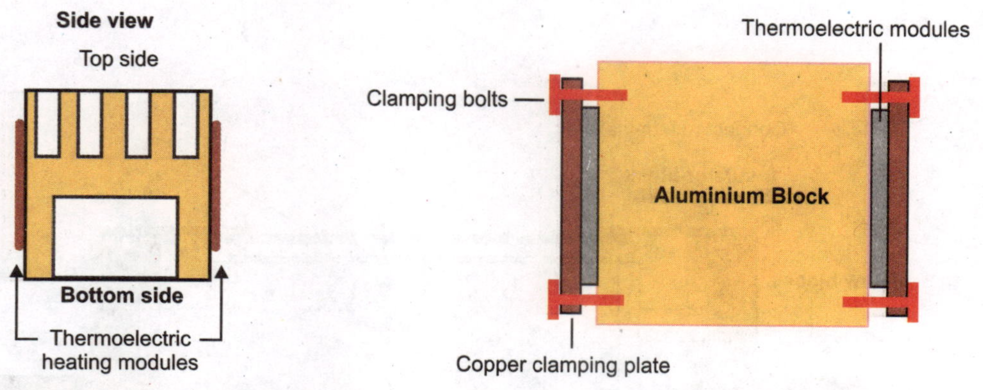

Fig. 34.2: Side view of metallic block

Fig. 34.3: Clamping of heating modules

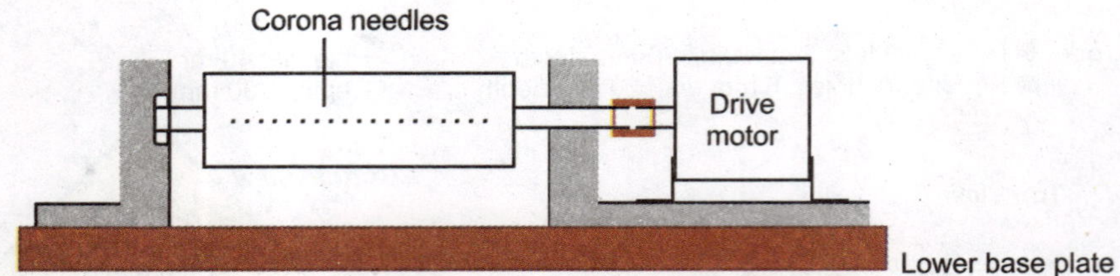

Fig. 37.4: Lower belt drive system

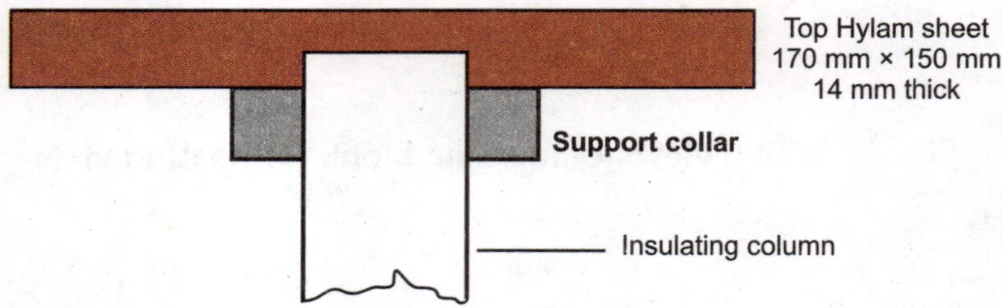

Fig. 37.5: Upper supporting platform in sulating column

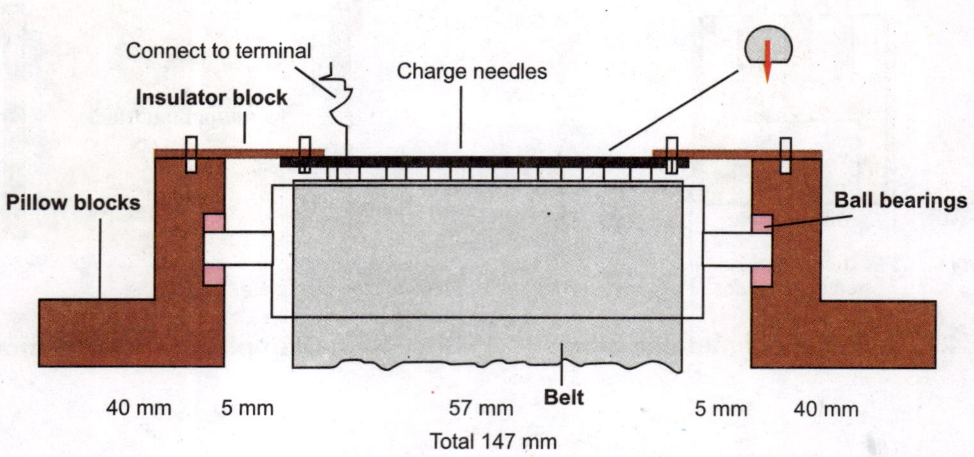

Fig. 37.6: Upper pulley assembly with charging needles